Mikiharu Kamachi · Akira Nakamura (Eds.)

New Macromolecular Architecture and Functions

Mikiharu Kamachi · Akira Nakamura (Eds.)

New Macromolecular Architecture and Functions

Proceedings of the OUMS '95 Toyonaka, Osaka, Japan, 2-5 June, 1995

With 136 Figures and 28 Tables

 Springer

Professor Mikiharu Kamachi
Professor Akira Nakamura
Osaka University
Dept. of Macromolecular Science
Toyonaka
560 Osaka
Japan

Library of Congress Cataloging-in-Publication Data
OUMS '95 (1995: Osaka, Japan)New macromolecular architecture and functions:
proceedings of the OUMS '95Toyonaka, Osaka, Japan, 2-5 June 1995 /
 Mikiharu Kamachi, Akira Nakamura(eds.).
Includes bibliographical references.
ISBN-13:978-3-642-80291-1   e-ISBN-13:978-3-642-80289-8
DOI: 10.1007/978-3-642-80289-8
1. Macromolecules--Congresses. 2. Polymerization--Congresses. 3. Polymers--Congresses.
I. Kamachi, Mikiharu. II. Nakamura, Akira. III. Title.
QD380.O94 1995
547.7--dc20   96-43179

ISBN-13:978-3-642-80291-1

Production: PRODUserv Springer Produktions-Gesellschaft, Berlin
Cover-Layout: de'blik, Berlin
Typesetting: Camera-ready copies from authors

SPIN: 10503791   2/3020 - 5 4 3 2 1 0 - Printed of acid-free paper

Springer

Berlin
Heidelberg
New York
Barcelona
Budapest
Hong Kong
London
Milan
Paris
Santa Clara
Singapore
Tokyo

Preface

This volume summarizes the papers presented at the second Osaka University Macromolecular Symposium OUMS '95 on "New Macromolecular Architecture and Functions" which was held at Senri Life Science Center, Osaka, Japan, on June 2 through June 5, 1995. The symposium covered the three topics, (1) Controlled Polymerizations, (2) Macromolecular Organized Systems and (3) Biomimetic Polymers, and invited leading scientists in these fields. At present, any of these topics is a hot issue in itself and frequently taken up separately on many occasions. It is noted, however, that all these topics are correlated with each other with the keyword "molecular design of new types of polymers" and their combination provides a unique feature of the present symposium in reflecting the interactions among investigators working in these important fields with the common ground expressed by the keyword "molecular design of new types of polymers". Twenty five invited lectures and twenty nine posters were presented at the Symposium, and twenty of the lectures contribute to this volume.

In the first topic, preparations of sequentially of stereoregularly controlled polymers were discussed from the view points of precise design of polymer preparation on the molecular level; attention was paid to a possibility of living radical polymerization, preparations of new types of living polymers, recent advances in preparation of stereospecific living polymers, sequential control in block copolymers, and molecular design of initiators and/or catalysts for the controlled polymerizations. The second topic was mainly concerned with preparation of supramolecular polymers and macromolecular organized systems, and their function. Preparations, structures, and functions of supramolecular polymers and macromolecular organized systems were discussed from the points of views of molecular recognition and assembly formation, and, in addition, molecular design of new types of materials. In the final topic, recent advances in preparations and fuctionalities of biomimetic polymers was shown, and characteristics of the polymers were discussed in relation to biological behavior and the preparation of new materials. Since the three topics are deeply correlated each other, the program of the symposium was arranged so that the three topics were properly mixed. The papers contributed to this volume should also reflect the outcome from such interaction at this meeting. Thus, we believe this volume will be useful for specialists working in these related fields on one hand, and serve as a refernce book for those who wish to get familiar with these fields on the other hand.

March 1996

Mikiharu Kamachi
Akira Nakamura

Table of Contents

Atom Transfer Radical Polymerization Including Degenerative Transfer: Novel and General Pathways Towards "Living" / Controlled Radical Polymerization

Scott G. Gaynor, Dorota Greszta, Jin-Shan Wang, Krzysztof Matyjaszewski*

Department of Chemistry, Carnegie Mellon University

4400 Fifth Avenue, Pittsburgh, PA 15213

ABSTRACT. A novel and general "living" radical polymerization, namely, Atom Transfer Radical Polymerization, ATRP, affords various pathways towards the synthesis of tailor - made (co)polymers with predetermined molecular weights up to $M_n \approx 10^5$ and molecular weight distributions as narrow as 1.05. This paper describes four different approaches to "living" ATRP, i.e., iodide-based degenerative transfer, alkyl halide / transition metal species promoted halogen atom transfer, $R\text{-}X / M_t^n / L_x$; radical initiator / transition metal species promoted halogen atom transfer, $I\text{-}I / M_t^{n+1} / L_x$; and alkyl halide / radical initiator / transition metal species promoted halogen atom transfer, $R\text{-}X / I\text{-}I / M_t^{n+1} / L_x$.

Introduction. Atom transfer radical addition, ATRA, is an important method for carbon - carbon bond formation in organic synthesis.[1] Two types of atom transfer methods have been developed. One of them is based on a univalent atom (e.g., hydrogen and halogen) or a group (e.g., SePh) transfer from a neutral molecule to a radical to form a new σ-bond and a new radical, **Scheme 1**.[2]

$$R\text{-}X \; + \; A^{\bullet} \; \xrightleftharpoons{\quad} \; R^{\bullet} \; + \; A\text{-}X$$

Scheme 1   A univalent atom or a group transfer radical reaction

Another atom transfer method is promoted by a transition metal species, in which the catalytic amount of the transition metal compound acts as a carrier of the halogen atom in a redox process, **Scheme 2**.[3] In analogy with ATRA, atom transfer radical polymerization, ATRP, can also be catalyzed by radicals or transition metal complexes.[4,5]

For successful atom transfer radical reactions, ATRA and ATRP both require the presence of a low concentration of free radicals in order to minimize the extent of termination reactions between radicals. However, there appears to be two principal differences between ATRP and ATRA.

M. Kamachi · A. Nakamura (Eds)

New Macromolecular Architecture and Functions

Proceedings of the OUMS '95 Toyonaka, Osaka, Japan, 2-5 June, 1995

© Springer-Verlag Berlin Heidelberg 1996

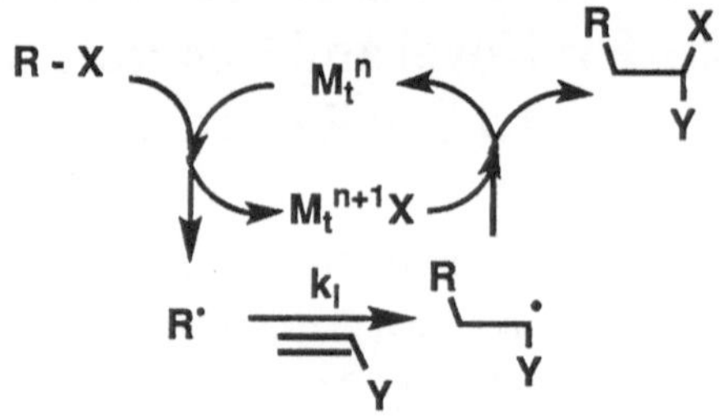

Scheme 2   Transition metal catalyzed atom transfer radical addition

First, an efficient ATRA process should avoid any kind of consecutive atom transfer processes, i.e., oligomerization or polymerization. This is the reason why, in ATRA, a very active atom transfer precursor and an inactive terminal alkene are often chosen by organic chemists. By contrast, ATRP involves a number of consecutive ATRA processes. Second, in ATRA most transfer reactions are irreversible under normal conditions, i.e., leading to degradative transfer, whereas the presence of fast and reversible atom transfer is necessary for obtaining well - defined polymers with predetermined molecular weights and low polydispersities.

In addition, it is well known that telomers can be synthesized by radical polymerization using either a transition metal species / alkyl halide combined initiating system or a radical initiator in the presence of a large amount of organic halide.[6] As discussed previously, in telomerization, the polymeric halides behave as dead chains.[5] They cannot be further activated, i.e., the transfer process is degradative (irreversible). Contrary to telomerization, ATRP represents a reversible transfer, in which the resulting polymeric halides are dormant species and can be repeatedly activated. Consequently, the molecular weight of the obtained polymer does not increase with monomer conversion in telomerization, whereas the molecular weight increases linearly with increasing monomer conversion in ATRP.

Degenerative Transfer. The detailed mechanism of a univalent atom X transfer based ATRP is outlined in **Scheme 3**. The radical, R·, is generated by the atom X transfer from R - X to an initiator radical, In·, which is generated by decomposition of a conventional initiator, e. g., AIBN or BPO. The radical R·, can then react with an alkene to form the radical species, R - M·. This species can subsequently abstract the X from R - X, to form a dormant species, R - M - X, and R·. The transfer process then occurs again. Because this process is a chain reaction, only small quantities of the initiator, relative to the transfer agent, are required to drive the reaction to completion. Assuming the reactivities of R - X and R - M - X are similar, competitive atom transfer may occur leading to the formation of polymers. Since the propagation in such a type of ATRP accompanies a thermodynamically neutral (reversible) exchange process, we called such a process *degenerative transfer*. A similar process probably also occurs in silyl ketene acetal-mediated group transfer polymerization in the presence of nucleophilic catalysts.[7]

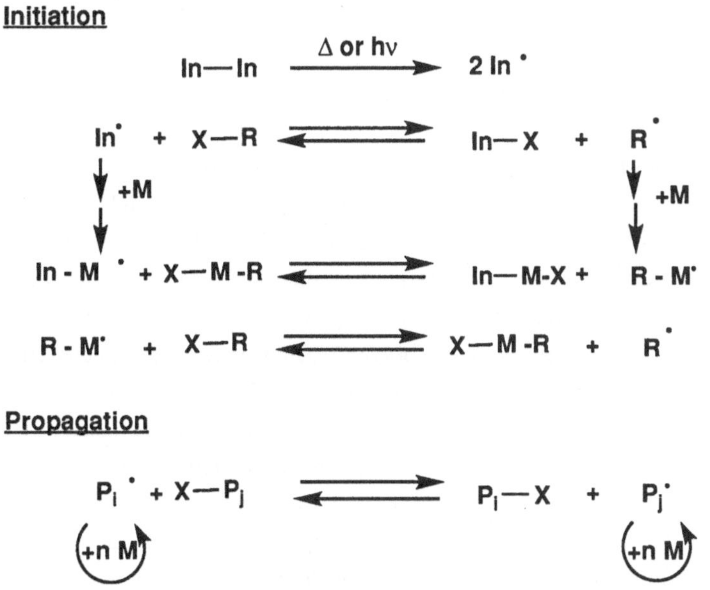

Initiation

$$In{-}In \xrightarrow{\Delta \text{ or } h\nu} 2\ In^\bullet$$

$$In^\bullet\ +\ X{-}R \rightleftharpoons In{-}X\ +\ R^\bullet$$

$$\downarrow +M \qquad\qquad\qquad\qquad\qquad \downarrow +M$$

$$In\text{-}M^\bullet\ +\ X{-}M\text{-}R \rightleftharpoons In{-}M\text{-}X\ +\ R\text{-}M^\bullet$$

$$R\text{-}M^\bullet\ +\ X{-}R \rightleftharpoons X{-}M\text{-}R\ +\ R^\bullet$$

Propagation

$$P_i^\bullet\ +\ X{-}P_j \rightleftharpoons P_i{-}X\ +\ P_j^\bullet$$

$$(+n\ M) \qquad\qquad\qquad\qquad\qquad (+n\ M)$$

Scheme 3
Atom transfer radical polymerization involving degenerative transfer

Successful degenerative transfer largely depends on how easily the radicals $R^\bullet$ and $P^\bullet$ can abstract the atom X from R - X and P - X, respectively. This is in turn determined by the structural nature of alkyl group, R and P, and transfer atom X.

Table 1 lists the dependence of bond dissociation energy, BDE, on atom X associated with benzyl radical.[8]

Table 1
Bond Dissociation Energies, BDE, of Various Benzyl Halides, $C_6H_5CH_2X$

X	I	Br	Cl	F
BED (kcal/mol)	48	58	73	>88

The following decreasing order of BDE is observed: R - I < R - Br < R - Cl < R - F. This is consistent with the decreasing reactivity of the reaction of a carbon - centered radical with R - X: R - I > R - Br > R - Cl >> R - F.[8] The range of reactivity is very large; alkyl iodides are outstanding atom donors due to a very weak R - I bond, i.e., lowest BDE. This explains why a controlled radical polymerization of styrene has been obtained using 1 - phenylethyl iodide as an initiator in the presence of a catalytic amount of AIBN, whereas 1 - phenylethyl bromide / AIBN combined initiating system affords only a conventional radical polymerization.[4] It is clear that bromine is not an efficient leaving group in degenerative transfer reactions.

The stability of the generated radicals $R^\bullet$ and $P^\bullet$ is very important in driving the direction and the rate of the atom transfer reaction. It has been demonstrated that the substituents containing resonance or inductive stabilizing group can reduce the BDE of the R - X and subsequently facilitate the atom transfer reaction.[2,4d]

In this regard, two points need to be addressed in relation to polymerization characteristics. First, as with nucleophilic substitution reactions, the energy gained by the forming P - X bond helps to lower the transition state energy to that required for complete cleavage of the R - X. Thus, the choice of the initiator, R - X, is crucial in obtaining the well-defined polymers with a predetermined M_n and narrow MWD. It was found that for an efficient degenerative transfer it is necessary to use R- X of the same structure as the dormant polymer chain end, e.g., 1-phenylethyl iodide in styrene polymerization, or with higher resonance or inductive stabilization energy than the polymer chain end, e.g., 1-phenylethyl iodide or iodoacetonitrile.[4d] Moreover, simple alkyl iodides such as butyl iodide were found to be inefficient degenerative transfer precursors, since the R - I bond in these transfer agents is strong, the atom transfer reaction is much slower in the initiation step than in the propagation one. Secondly, even though the R - I bond is very weak, so as to induce a fast atom transfer reaction in the initiation step, the stable P-I may result in a degradative transfer instead of degenerative transfer. This might be partially responsible for the observation that in the presence of 1-phenylethyl iodide, vinyl acetate does not undergo ATRP and no polymer was obtained, in contrast to the styrene and acrylate polymerization under the same conditions.

ATRP in the presence of transition metal complexes. (a) R - X / M_t^n /L_x combined initiating systems. Although a halide atom X, X = Cl and Br, in alkyl bromide and alkyl chloride can hardly be abstracted by free radical species, it can react with a number of transition metal species, M_t^n, resulting in the formation of radicals and the corresponding oxidized transition metal species, $M_t^{n+1}X$. If the reversible conversion of radicals to organic halide by reacting with $M_t^{n+1}X$ is fast and quantitative, a controlled atom transfer radical polymerization is observed. In this respect, the transition metal species acts as a carrier of atom X in a redox process between M_t^n and XM_t^{n+1} (**Scheme 4**). It has been demonstrated that the Cu^I/Cu^{II} redox cycle can induce a "living" ATRP of several alkenes.

Compared with other effective systems developed up to date[9], transition metal - promoted ATRP provides a more general and efficient method towards "living" radical polymerization. Several advantages are summarized as follows:

1) various commercially available and inexpensive organic halides can be used as efficient mono-, di-, and multi- functional initiators;

2) ATRP works well with a broad class of monomers, in principle, it could be applied to all radically polymerizable monomers[5b];

3) it can conceivably be carried out in bulk, solution, suspension, emulsion, and even in supercritical CO_2,

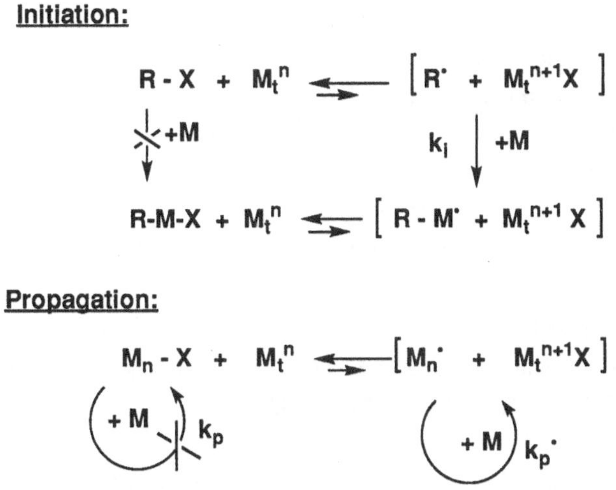

Initiation:

$$R - X + M_t^n \rightleftharpoons \left[R^\bullet + M_t^{n+1}X \right]$$

$$\downarrow +M \qquad\qquad\qquad k_i \,\Big|\, +M$$

$$R\text{-}M\text{-}X + M_t^n \rightleftharpoons \left[R\text{-}M^\bullet + M_t^{n+1}X \right]$$

Propagation:

$$M_n - X + M_t^n \rightleftharpoons \left[M_n^\bullet + M_t^{n+1}X \right]$$

$$\left(+M \right) k_p \qquad\qquad \left(+M \right) k_p^\bullet$$

Scheme 4

Transition metal catalyzed ATRP using R-X / M_t^n / L_x as initiating system

4) various well-defined (co)polymers with different topologies such as end-functional, block, random, tapered/gradient, star, comb/ grafting, hyperbranched, etc. can be, and have been, produced under moderate experimental conditions and with high efficiency, i.e., predetermined M_n up to 10^5 and M_w / M_n as low as 1.05.

Another very important feature of ATRP is that one may both thermodynamically and kinetically adjust the equilibration between dormant species and growing radicals, a key parameter for the control of the "living" radical polymerization. Use of different transition metal species, M_t^n, ligand(s), L_x, and transfer atom affects the "living" course of ATRP. On the other hand, since a catalytic amount of M_t^n / L_x is sufficient to promote ATRP, the "living" process in ATRP can also be controlled by varying the amount of the catalyst used. This makes the "living" ATRP more economic and practical.

The participation of free radicals in ATRP has been already evidenced by the observation of identical stereochemistry of both ATRP and conventional radical polymerization of MMA . Moreover, addition of 1.5 equiv. of galvinoxyl relative to alkyl chloride effectively inhibited the ATRP of styrene and no polymer was obtained after 18 hours.

Table 2 reports the monomer reactivity ratios, r, in ATRP and conventional random copolymerization of several co - monomer pairs. They are rather comparable, indicating a similar mechanism involved in two types of radical polymerization. Moreover, there appeared almost identical ^{13}C and ^{1}H NMR spectra, i.e., the same monomer sequence distribution in the random copolymers of styrene and MMA synthesized with AIBN / TEMPO - and ethyl 2 -bromoisobutyrate, Br - EIB, / CuBr / 2, 2' - bispyridine, bpy, based initiating systems under the same experimental conditions, i.e., in bulk and at 120 °C.

Table 2

Comparison of Monomer Reactivity Ratios of ATRP and Conventional Random Copolymerizations of Several Co - monomers

M_1	M_2	r_1 / r_2 (present work)	r_1 / r_2 (literature data)
Styrene	MMA	0.41 / 0.32[a]	0.43 /0.42[b]
		0.40 / 0.30 [c]	
MA	MMA	0.24 / 2.90[a]	0.36 / 2.23[d]
		0.23 / 3.71[c]	

[a] using Finemann-Ross and inverted Finemann-Ross methods; [b] K.F.O'Driscoll, et al. *J. Polym. Sci., Polym. Chem. Ed.* , **1984**, 22, 2776; [c] using Kelen-Tudos method; [d]V.P.Zubov, et al. *J. Polym. Sci., A-1* , **1971**, 9, 833.

This further supports the idea that a radical process is at the heart of ATRP promoted by CuBr / bpy. However, in comparison with the copolymer of MMA/St synthesized using a conventional initiator, dicumyl peroxide, DCPO, under the same experimental conditions, a small difference in ^{13}C NMR spectra of α–methyl group for MMA unit was observed

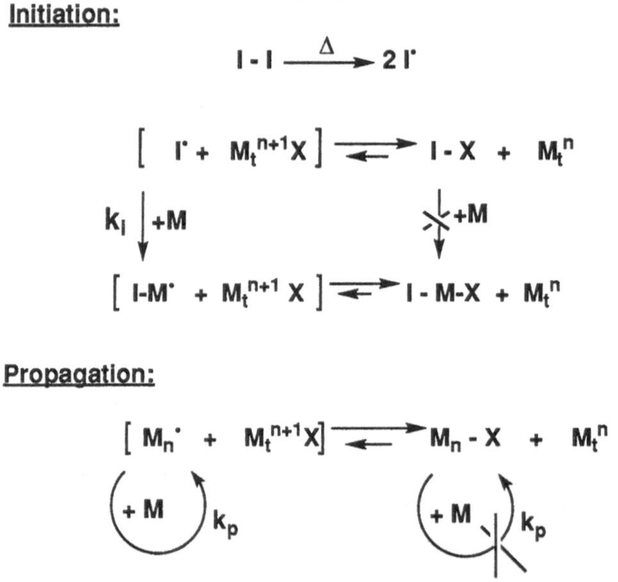

Initiation:

$$\text{I-I} \xrightarrow{\Delta} 2\,\text{I}^{\bullet}$$

$$\left[\ \text{I}^{\bullet} + \text{M}_t^{n+1}\text{X}\ \right] \rightleftharpoons \text{I-X} + \text{M}_t^n$$

$$k_i \downarrow +M \qquad\qquad \times\!\!\downarrow +M$$

$$\left[\ \text{I-M}^{\bullet} + \text{M}_t^{n+1}\,\text{X}\ \right] \rightleftharpoons \text{I-M-X} + \text{M}_t^n$$

Propagation:

$$\left[\ \text{M}_n^{\bullet} + \text{M}_t^{n+1}\text{X}\right] \rightleftharpoons \text{M}_n\text{-X} + \text{M}_t^n$$

$$\left(+M\right)k_p \qquad\qquad \left(+M\right)k_p$$

Scheme 5

Transition metal catalyzed ATRP using I-I / M_t^{n+1} / L_x initiating system

(b) **I-I / XM_t^{n+1} / L_x combined initiating systems.** It has been also demonstrated that a "living" ATRP of styrene can be obtained by using a conventional initiator, I - I, and in the presence of a higher oxidized transition metal complex XM_t^{n+1} / L_x (**Scheme 5**).

The mechanism is almost the same as R - X / M_t^n /L_x based ATRP , except at early stages in the initiation step. As an example, it has been already demonstrated that a controlled bulk ATRP of St can be obtained,[5d] using AIBN as an initiator and $CuCl_2$ (10 molar equiv.) / Bpy (20 molar equiv.) as a catalyst at 130°C. With increasing monomer conversion, M_n linearly increases and MWD was found to be as low as 1.30 . A linear semilogrithmic kinetic plot was also obtained. These results suggest a "living" ATRP process. This can be explained by very rapid consumption of AIBN, at 130 °C, to produce radicals which react with M_t^{n+1} - X / L_x to form R - X and M_t^n. These are the same species which were used in the above "living" system.

(c) **I - I / R - X / M_t^{n+1} /L_x combined initiating systems.** In comparison with R - X / M_t^n /L_x based ATRP, I - I / M_t^{n+1} / L_x based system usually requires a larger amount of M_t^{n+1} / L_x catalyst, since the concentration of free radicals generated from radical initiator I - I is higher. Moreover, the reactivity of radical $P^•$ towards M_t^{n+1} and solubility of M_t^{n+1} / L_x species are also critical in order to obtain a "living" ATRP prior to a conventional radical polymerization. For example, although St bulk ATRP results in a predetermined M_n and narrow MWD (M_w / M_n ~ 1.3) using AIBN as an initiator and in the presence of 10 molar equiv. of $CuCl_2$ and 20 molar equiv. of bpy, MA polymerization under the same conditions leads to an ill - controlled process (Fig. 1).

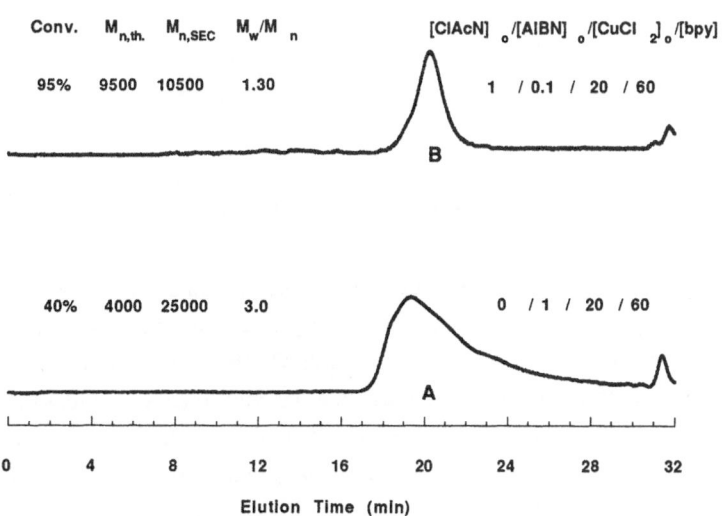

Figure 1. Comparison of SEC traces of bulk MA polymerization using AIBN / $CuCl_2$ / bpy initiating system at 130 °C with (B) and without (A) 2 - chloroacetonitrile, 2 - ClACN.

The mechanism of R-X / I-I / M_t^{n+1} / L_x promoted ATRP is illustrated in Scheme 6.

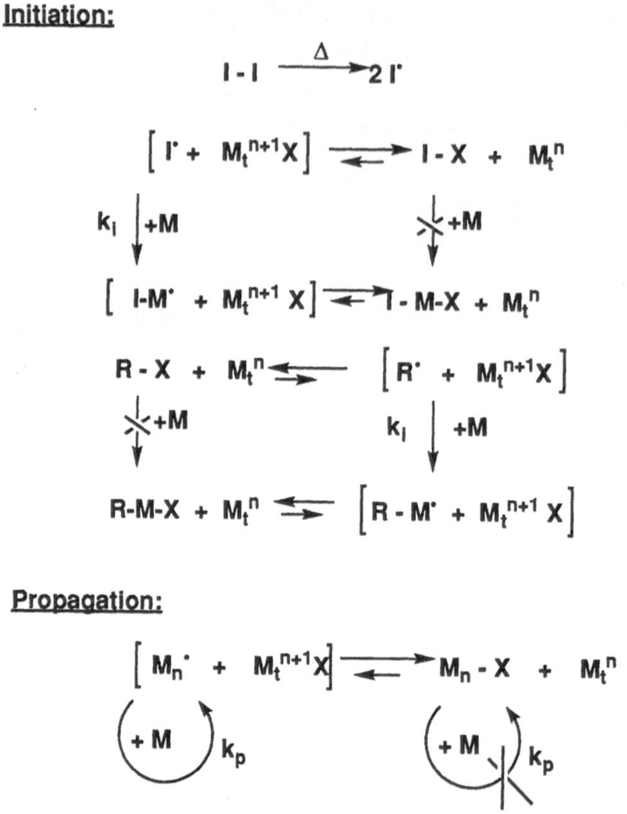

Initiation:

$$I-I \xrightarrow{\Delta} 2\ I^{\cdot}$$

$$\left[I^{\cdot} + M_t^{n+1}X \right] \rightleftarrows I-X + M_t^n$$

$k_i \downarrow +M \qquad\qquad \nmid +M$

$$\left[I-M^{\cdot} + M_t^{n+1}X \right] \rightleftarrows I-M-X + M_t^n$$

$$R-X + M_t^n \rightleftarrows \left[R^{\cdot} + M_t^{n+1}X \right]$$

$\nmid +M \qquad\qquad k_i \downarrow +M$

$$R-M-X + M_t^n \rightleftarrows \left[R-M^{\cdot} + M_t^{n+1}X \right]$$

Propagation:

$$\left[M_n^{\cdot} + M_t^{n+1}X \right] \rightleftarrows M_n-X + M_t^n$$

$\left(+M \right) k_p \qquad\qquad \left(+M \right) k_p$

Scheme 6

Transition metal catalyzed ATRP using R-X(I-I) / M_t^{n+1} / L_x initiating system

As seen in **Figure 1A**, the SEC of THF soluble portion shows a broad distribution with Mw / Mn ≈ 3.0. This has been explained by the presence of high concentration of growing radicals and of lower reactivity of PMA• towards CuCl₂ / bpy as compared to PSt•. Such a problem can be solved simply by using AIBN in a catalytic amount. In order to simultaneously control the molecular weight of the polymer, an alkyl halide, R-X,, e.g., 2-chloroacetonitrile, 2-ClACN, has to be used. In this regard, the molecular weight can be predetermined by means of $DP_n = [M]_0 / ([R-X]_0 + 2[AIBN]_0)$. As seen in **Figure 1B**, ATRP of MA is a controlled process, i.e., high monomer conversion (>90%), predetermined M_n, and narrow molecular weight distribution (M_w/M_n ~ 1.30), using a catalytic amount of AIBN (up to 10%) and initiator amount of 2 - chloroacetonitrile in the presence of 10 molar equiv. of CuCl₂ and 20 molar equiv. of bpy (both relative to 2 - chloroacetonitrile) at 130 °C.

Conclusions

A novel and general "living" radical polymerization, namely atom transfer radical polymerization, ATRP, provides for a simple approach to the synthesis of tailor - made (co)polymers with predetermined molecular weight, low polydispersities and novel topologies.

References.

(1) Curran, D. P. *Synthesis,* **1988,** 489.

(2) Curran, D. P. in *Comprehensive Organic Synthesis,* Trost, B. M., Fleming, I., Eds., Pergamon: Oxford, 1991, Vol. 4, p 715, and reference therein.

(3) (a) Nagashima, H.; Wakamatsu, H.; Ozaki, N., Ishii, T.; Watanabe, M.; Tajima, T.; Itoh, K. *J. Org. Chem.* **1992,** *57,* 1682; (b) Udding, J. H.; Tuijp, K. J. M.; van Zanden, M. N. A.; Hiemstra, H.; Speckamp, W. N. *J. Org. Chem.* **1994,** *59,* 1993.

(4) (a) Matyjaszewski, K.; Gaynor, S.; Wang, J.S. *Macromolecules,* **1995,** 28, 2093; (b) Gaynor, S.; Wang, J.S.; Matyjaszewski, K. *Polym. Prep.* (Am. Chem. Soc., Polym. Chem. Div.), **1995,** 36(1), 467; (c) Wang, J.S.; Gaynor, S.; Matyjaszewski, K. *Polym. Prep.* (Am. Chem. Soc., Polym. Chem. Div.), **1995,** 36(1), 465; (d) Matyjaszewski, K.; Gaynor, S.; Wang, J.S. *Macromolecules,* submitted

(5) (a) Wang, J.S.; Matyjaszewski, K. *J. Am. Chem. Soc.* **1995,** 117, 5614; (b) Matyjaszwski, K.; Wang, J.S. *U. S. Pat.* pending; (c) Wang, J.S.; Matyjaszewski, K. *Macromolecules,* in press; (d) Wang, J.S.; Matyjaszewski, K. *Macromolecules,* submitted.

(6) (a) Boutevin, B.; Pietrasanta, Y. In *Comprehensive Polymer Science,* Allen, G.;, Aggarwal, S. L.; Russo, S. Eds., Pergamon: Oxford, 1991, vol. 3, p 185; (b) Bamford, C. H. In *Comprehensive Polymer Science (First Supplement),* Allen, G.; Aggarwal, S. L.; Russo, S. Eds., Pergamon: Oxford, 1992, p 1.

(7) Muller, A.H.E.; Zhuang, R.; Yan, D.; Litvinenko, G. *Macromolecules* **1995,** 28, 4326

(8) (a) Danen, W. C., in *Methods in Free Radical Chemistry,* Huyser, E. L. S. Ed., Dekker, New York, 1974, vol. 5, p.1; (b) Poutsma, M. in *Free Radicals,* Kochi, J.K. Ed., Wiley, New York, 1973, vol. 2, p.113.

(9) (a) Otsu, T.; Yoshida, M. *Makromol. Chem., Rapid Commun.* **1982,** 3, 127, 133; (b) Solomon, D. H.; Rizzardo, E.; Cacioli, P.; U.S. Patent 4,581,429, March 27, 1985. (c) Georges, M.K.; Veregin, R.P.N.; Kazmaier, P.M.; Hamer, G. K. *Macromolecules* **1993,** 26, 2987; (d) Wayland, B. B.; Poszmik, G.; Mukerjee, S.L.; Fryd, M. *J. Am. Chem. Soc.* **1994,** 116, 7943.

Living Radical Polymerization via Reversible Homolytic Activation of Carbon-Halogen Bonds with Metal Complexes

Mitsuo Sawamoto* and Masami Kamigaito

Department of Polymer Chemistry, Kyoto University
Kyoto 606-01, Japan

Abstract: Living radical polymerizations of methyl methacrylate and related acrylic monomers have been achieved with ternary initiating systems that consist of an alkyl chloride (R–Cl), ruthenium(II) chloride tris(triphenylphosphine) complex [RuCl2(PPh3)3], and a Lewis acid (MX$_n$), where R–Cl = CCl4, CCl3CO2-CH3, CCl3COCH3, CCl3COC6H5, etc.; MX$_n$ = AlCH3(2,6-di-*t*-butylphenoxy)2, Al(O*i*Pr)3, Ti(O*i*Pr)4, etc. The produced polymers had controlled molecular weights, nearly proportional to conversion, and narrow molecular weight distributions; in particular, the CCl3COCH3/RuCl2(PPh3)3/Al(O*i*Pr)3 system led to polymers with very narrow MWDs ($M_w/M_n \leq 1.1$). Sequential polymerizations of methacrylic and acrylic monomers also gave AB- and ABA-type block copolymers.

INTRODUCTION

Living Polymerizations towards Controlled Macromolecular Architectures

In the frontiers in the contemporary polymer chemistry certainly lies design and development of precision polymerizations that give well-defined and controlled macromolecular architectures, as extensively discussed during this Symposium in Osaka. Living polymerizations are perhaps the best in such processes in addition polymerizations, and in the 1980s and 1990s a number of new living polymerization reactions have been developed in virtually all but radical propagation mechanisms [1]. For example, we have been developing living cationic polymerization of vinyl monomers, initiated with protonic acid/metal halide binary initiating systems [2] (Scheme 1). Via these processes, it is now feasible to synthesize a variety of well-defined polymers with specific functionalities.

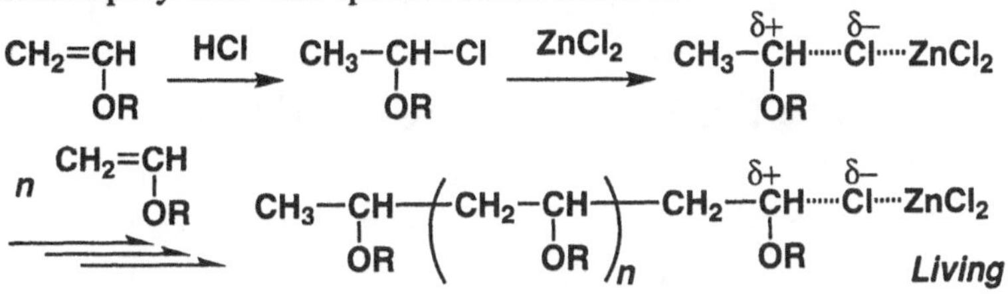

Scheme 1. Living cationic polymerization of vinyl ethers with HCl/ZnCl2 initiating system [3].

M. Kamachi · A. Nakamura (Eds)
New Macromolecular Architecture and Functions
Proceedings of the OUMS '95 Toyonaka, Osaka, Japan, 2-5 June, 1995
© Springer-Verlag Berlin Heidelberg 1996

As illustrated in Scheme 1 for the polymerization with the HCl/ZnCl$_2$ system [3], we have shown that a reversible and dynamic equilibrium between growth-active intermediate and its dormant counterpart plays a critical role in controlling the propagation process and in eliminating undesirable side reactions such as chain transfer [3,4]. Namely, hydrogen chloride (initiator) generates the covalent carbon–chlorine bond in a polymer terminal that is dormant and per se incapable of propagating. The dormant end is, however, activated or ionized by zinc chloride (activator or Lewis acid) into a carbocationic species that is the true growing species. These *dormant* and *active* species are also shown to be in a rapid and reversible interchange equilibrium, which in turn decreases the instantaneous concentration of the active form and thereby suppresses chain transfer and other side reactions. The rapid interchange also ensures the virtually equal probability of growth for all dormant species to give uniform molecular weights or narrow molecular weight distributions (MWDs).

The interchange equilibrium thus involves a reversible and heterolytic cleavage of a carbon–halogen terminal bond mediated by a Lewis acid. The general importance of such reversible equilibria in controlling growth processes appear to be verified in a variety of cationic and other living polymerizations [1,4].

From Heterolytic to Homolytic Activation of Carbon–Halogen Bonds towards Living Radical Polymerizations

It then occurred to us that the Lewis acid-assisted *hetelolytic* (cationic) cleavage of carbon–halogen bonds might be extended, by analogy, to *homolytic* (radical) cleavage of similar linkages by which we might achieve *living radical* polymerizations. Scheme 2 conceptually compares the two processes that should be dynamic and *reversible*, as discussed above. To affect such homolytic and reversible cleavage, in our view, metal complexes will be critically required, as metal halides or Lewis acids do in the hetelolytic/cationic counterparts (Scheme 1).

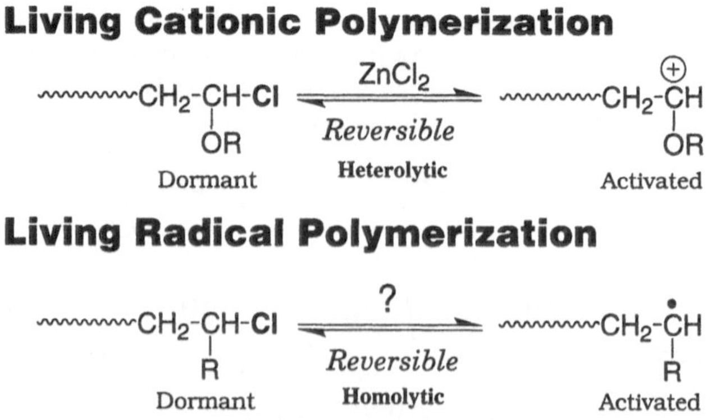

Living Cationic Polymerization

$$\text{\small\textasciitilde\textasciitilde\textasciitilde CH}_2\text{-CH-Cl} \underset{\textit{Reversible} \quad \text{Heterolytic}}{\overset{\text{ZnCl}_2}{\rightleftharpoons}} \text{\small\textasciitilde\textasciitilde\textasciitilde CH}_2\text{-}\overset{\oplus}{\text{CH}}$$

$$\quad\quad | \quad\quad\quad\quad\quad\quad\quad\quad\quad\quad\quad\quad\quad\quad | $$
$$\quad\quad \text{OR} \quad\quad\quad\quad\quad\quad\quad\quad\quad\quad\quad\quad\quad \text{OR}$$

Dormant   Activated

Living Radical Polymerization

$$\text{\small\textasciitilde\textasciitilde\textasciitilde CH}_2\text{-CH-Cl} \underset{\textit{Reversible} \quad \text{Homolytic}}{\overset{?}{\rightleftharpoons}} \text{\small\textasciitilde\textasciitilde\textasciitilde CH}_2\text{-}\overset{\bullet}{\text{CH}}$$

$$\quad\quad | \quad\quad\quad\quad\quad\quad\quad\quad\quad\quad\quad\quad\quad\quad | $$
$$\quad\quad \text{R} \quad\quad\quad\quad\quad\quad\quad\quad\quad\quad\quad\quad\quad \text{R}$$

Dormant   Activated

Scheme 2. Living polymerizations via reversible cleavage of carbon–halogen bonds.

Literature search has in fact revealed that a carbon–chlorine bond in carbon tetrachloride can be homolytically cleaved by dichlorotris(triphenylphosphine)ruthenium(II) [RuCl2(PPh3)3] to form the trichloromethyl radical, which undergoes addition to an alkene into an adduct (**1**) (eq 1) [5]. We theorized that subsequent homolytic and reversible activation of the carbon–chlorine bonds in **1** and its polymeric homologues might lead to propagation with radically polymerizable alkenes and then to living radical polymerizations that resemble mechanistically to our living cationic polymerizations.

$$CCl_4 \; + \; \underset{R}{CH_2{=}CH} \; \xrightarrow[\left[CCl_3{\cdot}\right]]{RuCl_2(PPh_3)_3} \; \underset{R}{CCl_3{-}CH_2{-}CH{-}Cl} \qquad (1)$$

$$\mathbf{1}$$

This article describes our development of living radical polymerization of methacrylates and acrylates based on the homolytic and reversible activation of terminal carbon–chlorine linkages assisted by the ruthenium complex [6–8].

LIVING RADICAL POLYMERIZATION WITH CCl4/RuCl2(PPh3)3 SYSTEMS

Living Polymerization of Methyl Methacrylate [6]

Following the above-mentioned strategy, methyl methacrylate (MMA) was polymerized with the combination of carbon tetrachloride and the Ru complex [CCl4/RuCl2(PPh3)3] in toluene solvent at 60–80 °C. Initial attempts, however, soon revealed that the binary system is totally incapable of polymerizing MMA under these conditions. In the additional presence of an aluminum Lewis acid, methylaluminum bis(2,6-di-t-butylphenoxide) [MeAl(ODBP)2], on the other hand, readily effected a smooth and quantitative polymerization that reached above 90 % conversion in 4 h: $[M]_0 = 2.0$ M; $[CCl_4]_0 = 20$ mM; $[RuCl_2(PPh_3)_3]_0 = 10$ mM; $[MeAl(ODBP)_2]_0 = 40$ mM. The logarithmic conversion ($\ln[M]_0/[M]_t$) is proportional to reaction time t, indicating the constancy of the active center concentration and most likely the absence of termination.

The ruthenium-mediated polymerization can also be continued by adding a fresh feed of MMA to a completely polymerized reaction system, which triggers a smooth second-phase polymerization at nearly the same rate as in the first. As shown in Figure 1, the polymers produced before and after the monomer addition exhibit fairly narrow MWDs ($M_w/M_n \leq 1.2$) that continuously shift towards higher molecular weights without broadening but even with narrowing. More important, the number-average molecular weights (M_n) increase almost linearly with conversion even after the monomer addition. These observations demonstrate that the ternary initiating system, CCl4/RuCl2(PPh3)3/MeAl(ODBP)2, in fact induces a new living polymerization of MMA.

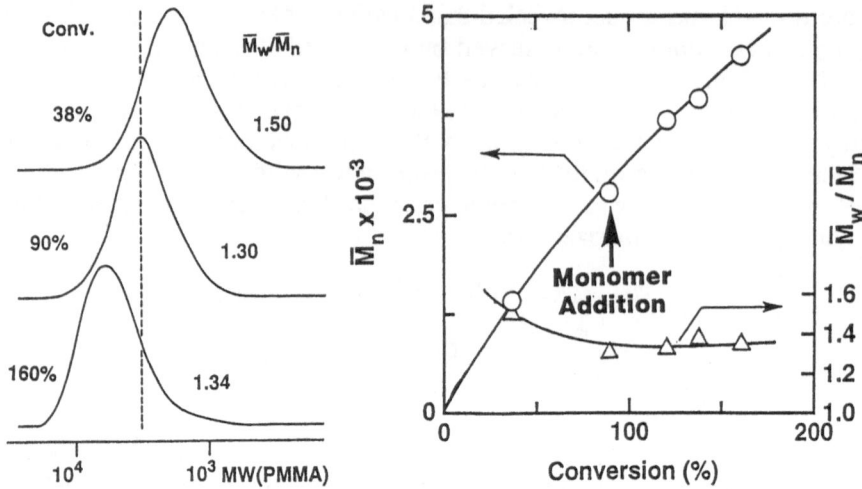

Figure 1. Living radical polymerization of MMA with CCl4/ RuCl2-(PPh3)3/MeAl(ODBP)2, (20/10/40 mM) in toluene at 60 °C: monomer addition experiments. $[M]_0 = [M]_{added} = 2.0$ M [6].

Polymerization Pathway

Separate experiments show that the three components of the initiating systems are indispensable for the living polymerization to occur; there is no consumption of MMA in the absence of one or more of the three under our reaction conditions. Note that the polymerization requires the aluminum Lewis acid [MeAl(ODBP)2] as the third component, which indicates that the initial MMA–CCl4 adduct (**1**; eq 1) per se cannot initiate polymerization in the absence of the Lewis acid. The polymer molecular weights are nearly inversely proportional to the initial concentration of CCl4, and this compound is quantitatively depleted in the very early stages of the polymerization. Thus, CCl4 most likely serves as an initiator.

As suggested in the CCl4 addition to alkenes (eq 1), the Ru-mediated MMA polymerization appears to proceed in a radical mechanism. Supporting evidence for this includes: (i) the complete suppression of the polymerization in the presence of radical inhibitors or scavengers (e.g., galvinoxyl); (ii) the syndiotactic-rich steric structure of the polymers where the S:H:I triad ratio (64:33:3) is very similar to that (62:35:3) for a poly(MMA) sample obtained with AIBN otherwise under the same conditions.

On the basis of these observations, we have proposed that the MMA living polymerization with the Ru system proceeds as in the pathway shown in Scheme 3 [6]. The polymerization may be initiated from the initial adduct **1** in the presence of RuCl2(PPh3)3 and MeAl(ODBP)2. The first propagation step may involve the

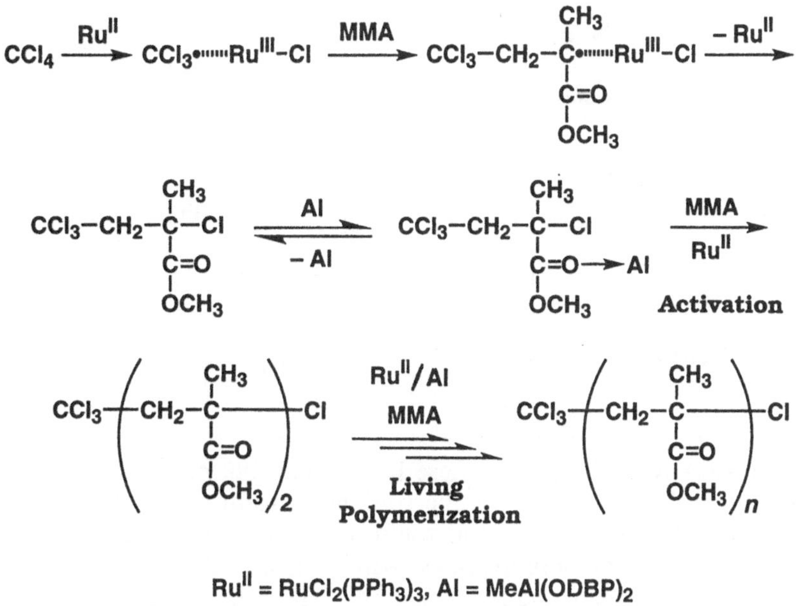

$$CCl_4 \xrightarrow{Ru^{II}} CCl_3\bullet\text{·····}Ru^{III}\text{-}Cl \xrightarrow{MMA} CCl_3\text{-}CH_2\text{-}\underset{\underset{OCH_3}{\overset{|}{\underset{|}{C=O}}}}{\overset{CH_3}{\overset{|}{C}}}\bullet\text{·····}Ru^{III}\text{-}Cl \xrightarrow{-\ Ru^{II}}$$

$$CCl_3\text{-}CH_2\text{-}\underset{\underset{OCH_3}{\overset{|}{\underset{|}{C=O}}}}{\overset{CH_3}{\overset{|}{C}}}\text{-}Cl \underset{-\ Al}{\overset{Al}{\rightleftharpoons}} CCl_3\text{-}CH_2\text{-}\underset{\underset{OCH_3}{\overset{|}{\underset{|}{C=O\rightarrow Al}}}}{\overset{CH_3}{\overset{|}{C}}}\text{-}Cl \underset{Ru^{II}}{\overset{MMA}{\longrightarrow}}$$

Activation

$$CCl_3\left(CH_2\text{-}\underset{\underset{OCH_3}{\overset{|}{\underset{|}{C=O}}}}{\overset{CH_3}{\overset{|}{C}}}\right)_2\text{-}Cl \underset{\underset{\text{Living}}{\underset{\text{Polymerization}}{}}}{\overset{Ru^{II}/Al}{\underset{MMA}{\rightleftharpoons}}} CCl_3\left(CH_2\text{-}\underset{\underset{OCH_3}{\overset{|}{\underset{|}{C=O}}}}{\overset{CH_3}{\overset{|}{C}}}\right)_n\text{-}Cl$$

$Ru^{II} = RuCl_2(PPh_3)_3$, $Al = MeAl(ODBP)_2$

Scheme 3. Living radical polymerization of MMA with CCl4/RuCl2-(PPh3)3/MeAl(ODBP)2: a proposed reaction pathway.

interaction of the Ru complex with the carbon–chlorine bond in **1** into a radical species; the aluminum Lewis acid may facilitate the homolytic cleavage of the bond into a transient radical, probably via its interaction with the ester moiety of the adduct (see below). Subsequently, successive additions of MMA monomers to these radical species in a similar fashion give a series of homologues of **1** that carry the carbon–chlorine terminals structurally identical to that in **1**. Very recently, a similar controlled polymerizations of MMA and related monomers have been reported, where the Ru complex is replaced with Cu complexes [9].

DESIGN OF INITIATING SYSTEMS

Lewis Acids [7]

A notable characteristic of the Ru-mediated living polymerization is the need of MeAl(ODBP)2 as the third component of the initiating system, which fact clearly differentiates it from the corresponding radical addition of CCl4 to alkenes previously reported (eq 1) [5]. To determine the scope of such components, polymerization experiments were carried out with the CCl4/RuCl2(PPh3)3 system in the presence of a series of metal alkoxides, phenoxides, and halides, listed below, that are considered as relatively weak Lewis acids: $[M]_0 = 2.0$ M; $[CCl4]_0 = 20$ mM; $[RuCl2(PPh3)3]_0 = 10$ mM; [Lewis acid]$_0 = 40$ mM; in toluene at 60 °C.

Similar to MeAl(ODBP)$_2$, all of these Lewis acids but metal halides (SnCl$_4$ and TiCl$_4$) induced living MMA polymerizations to form polymers with relatively narrow MWDs ($M_w/M_n \leq 1.2$) and controlled molecular weights. In particular, aluminum and titanium isopropoxides [Al(OiPr)$_3$ and Ti(OiPr)$_4$] are effective in terms of molecular weight and MWD control. Thus, the Lewis acids are generally important in these reactions, and bulky substituents, like the di-t-butyl in MeAl(ODBP)$_2$, are not necessarily needed to induce the living processes.

Although the exact role of these Lewis acids is an open question at present, they are expected to interact with, or coordinate onto, (i) the ester carbonyl of the dormant terminal with a carbon–chlorine bond, (ii) the ester carbonyl of the monomer, and/or (iii) the ruthenium center of RuCl$_2$(PPh$_3$)$_3$. In view of the absence of the polymerization with the CCl$_4$–MMA adduct 1 and RuCl$_2$(PPh$_3$)$_3$ alone, we incline to suppose the possibility of (i) where the coordination may activate the carbon–chlorine terminal bond so as to be homolytically cleaved in the presence of the Ru complex.

Initiators [8]

If the proposed reaction pathway (Scheme 3) is valid, not only CCl$_4$ but also compounds (R–Cl) with similar carbon–chlorine (and perhaps carbon–halogen) linkages may serve as initiators in the Ru-mediated living polymerization. Thus, the compounds (shown in the next page) were examined as possible initiators, which include esters, ketones, and alkyl and aryl halides.

In toluene at 60–80 °C, MMA was polymerized with R–Cl coupled with RuCl$_2$-(PPh$_3$)$_3$ and Al(OiPr)$_3$. All these initiating systems induced polymerizations, among which CCl$_3$CO$_2$CH$_3$, CHCl$_2$CO$_2$CH$_3$, and CCl$_3$COCH$_3$ were particularly effective. The polymerization rates with the three systems are nearly the same as that with CCl$_4$ under the same conditions, and the produced polymers invariably showed narrow MWDs.

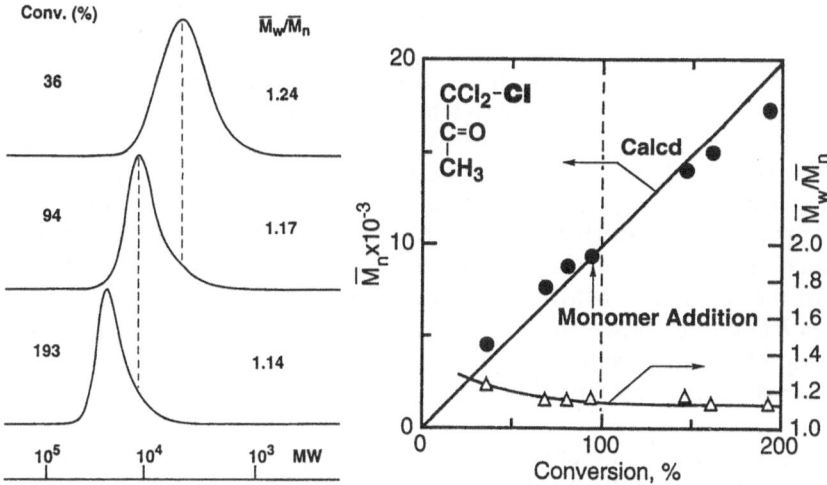

For example, Figure 2 shows the M_n and MWD of the polymers obtained with the CCl3COCH3/RuCl2(PPh3)3/Al(OiPr)3 system. The MWDs clearly narrowed as the polymerization proceeds to give very narrow distributions with M_w/M_n close to 1.1. The M_n is directly proportional to conversion, both before and after sequential addition of MMA, and is in close agreement with the calculated value assuming that each molecule of the ketone forms one polymer chain. Thus, truly living polymerization has been achieved with the CCl3COCH3-based initiating system to give poly(MMA) with controlled molecular weights and very narrow MWDs, perhaps for the first time in the metal-mediated radical living polymerizations.

Figure 2. Living radical polymerization of MMA with CCl3COCH3/ RuCl2(PPh3)3/Al(OiPr)3 (20/10/40 mM) in toluene at 80 °C: monomer addition experiments. [M]0 = [M]added = 2.0 M [8].

The effectiveness of these R–Cl type compounds strongly supports the proposed reaction pathway (Scheme 3), in which the cleavage of the carbon–chlorine bond is required in the initiation step. As readily perceived, the three initiators discussed above invariably have a carbonyl group attached to the α-carbon where the initiating carbon–chlorine bond is attached, and their common structure (–CO–C–Cl) is akin to the dormant polymer terminal expected from the mechanism.

MONOMERS AND BLOCK COPOLYMERS

Polymerization of Methacrylates and Acrylates [10]

The ruthenium-based initiating systems effective for MMA are also applicable to other alkyl methacrylates and acrylates. For example, in toluene at 80 °C the CCl4/RuCl2(PPh3)3/Al(OiPr)3 system effectively polymerize not only MMA but also ethyl and n-butyl methacrylates to form living polymers with fairly narrow MWDs ($M_w/M_n \leq 1.2$). Under these conditions, polymer molecular weights increase progressively with conversion. The corresponding acrylates can also be polymerized with the same and related initiating systems at similar rates, but polymer MWDs are broader.

Block Copolymerizations [10]

The expansion of the scope of monomers thus led to the synthesis of block copolymers by sequential living polymerization. Figure 3 illustrates typical examples, where di- and tri-block copolymers are synthesized from MMA and n-butyl methacrylate (nBMA).

Thus, with the CCl4/RuCl2(PPh3)3/Al(OiPr)3 initiating system, MMA is first polymerized into living polymers to which nBMA is added. The addition leads to a clear increase in molecular weight, as seen in the middle MWD trace, and gives MMA–nBMA AB-type block copolymers. Further addition of MMA to this polymer (unquenched) again leads to a third-phase polymerization where the MWD curve shifted clearly to higher molecular weights and thus resulted in ABA-type block copolymers of MMA and nBMA. Note, in particular, that there is no trace of the remaining MMA homopolymer and the AB block precursors in the final MWD curve shown in the bottom. These results further demonstrate that each step of the three sequential polymerizations is living, or virtually free from chain transfer and termination (polymer recombination, etc.).

In conclusion, the ruthenium-based new initiating systems have been developed that achieved living radical polymerizations of MMA and related monomers. We must admit that in these polymerizations, a number of points to be clarified in terms of the mechanism, roles of each component in the ternary initiating systems, etc. Extensive study is now in progress in our laboratories.

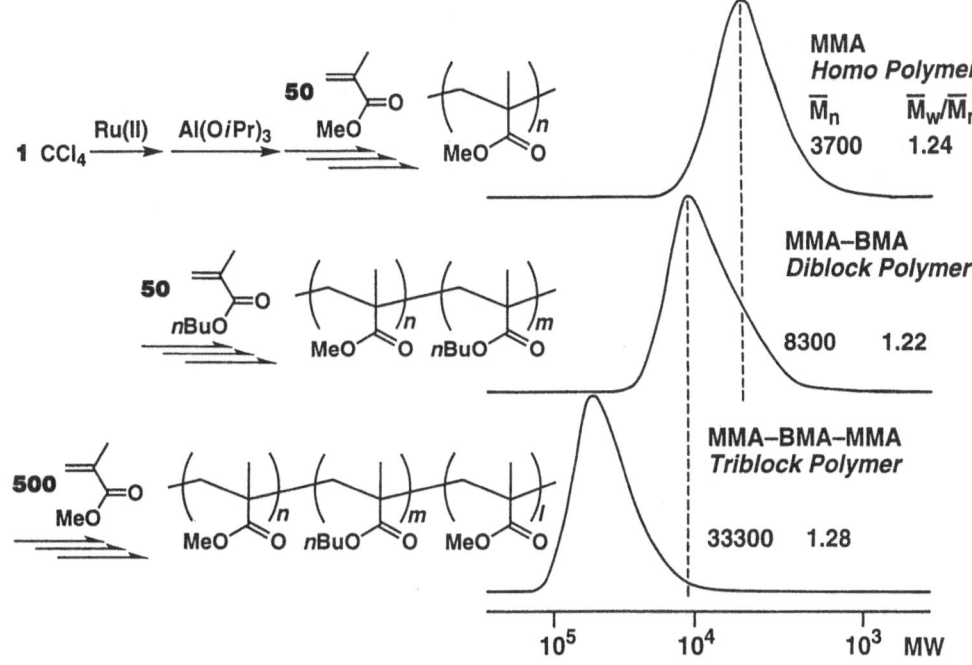

Figure 3. Block copolymerization of MMA and nBMA with CCl4/RuCl2(PPh3)3/Al(OiPr)3 (40/10/40 mM) in toluene at 80 °C.

Acknowledgments: The authors thank Professor Toshinobu Higashimura for his comments and Mitsuru Kato, Tuyoshi Ando, and Yuzo Kotani for their contributions to the work described in this article.

REFERENCES

1 Webster OW (1991) *Science* **251**: 887
2 Sawamoto M (1991) *Prog Polym Sci* **16**: 111; (1993) *Trends Polym Sci* **1**:111
3 Kamigaito M, Maeda Y, Sawamoto M, Higashimura T (1993) *Macromolecules* **26**: 2670
4 Katayama H, Kamigaito M, Sawamoto M, Higashimura T (1995) *Macromolecules* **28**: 3747; (1995) *J Phys Org Chem* **8**: 282
5 Matsumoto H, Nakano T, Nagai Y (1973) *Tetrahedron Lett* **51**: 5147
6 Kato M, Kamigaito M, Sawamoto M, Higashimura T (1995) *Macromolecules* **28**: 1721; (1994) *Polym Prepr Jpn* **43**:1792
7 Ando T, Kato M, Kamigaito M, Sawamoto M (1995) *Polym Prepr Jpn* **44**: 111

8 Ando T, Kato M, Kamigaito M, Sawamoto M (1995) *Polym Prepr Jpn* **44**: 110; *Macromolecules* submitted
9 Wang J-S, Matyjaszewski K (1995) *J Am Chem Soc* **117**: 5614
10 Kotani Y, Kato M, Kamigaito M, Sawamoto M (1995) *Polym Prepr Jpn* **44**: 110

Molecular Information and Its Expression by Oligomeric Charbon-Chain Compounds; What Steroidal Molecules Tell Us

Mikiji Miyata

Faculty of Engineering, Osaka University, Yamadaoka, Suita, Osaka 565, JAPAN

Abstract: We studied on molecular assemblies of steroids involving multiple hydrogen bonding groups which play diversible roles in determining their association, inclusion, recognition, intercalation and reaction. This study led us to a creation of the concept that even oligomeric carbon-chain compounds can express their molecular information through the assemblies.

INTRODUCTION

People consider that biopolymers, such as nucleic acids and proteins, have molecular information and express them. More generally, however, why don't they think that organic molecules do likewise ? In other words, what about some molecules other than the biopolymers ? It is noteworthy that hypothetical removal of amide linkages from proteins brings about carbon-chain compounds or substituted polymethylenes, as shown in Figure 1. During the study of inclusion polymerization by using steroids, called deoxycholic acid (Figure 2), I got the idea that steroids can be regarded as the polymethylenes suitable for the expression of molecular information.[1,2] Namely, even oligomeric carbon-chain compounds can express the information through their assemblies. Here I want to summarize a series of our study on molecular assemblies of steroids, starting from a consideration about the universe.

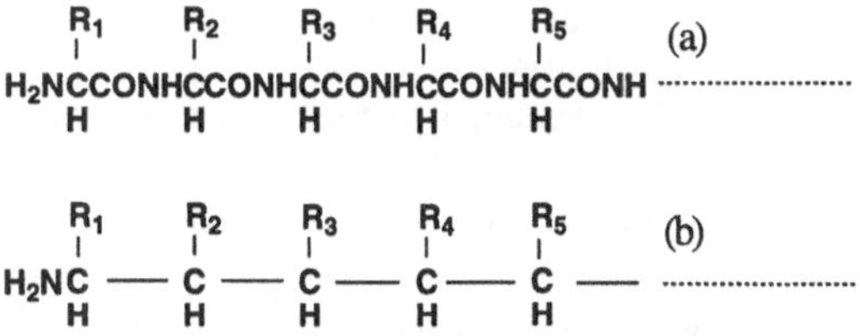

Figure 1 Sequential peptide chain (a) and carbon-chain (b) after removal of its amide linkages.

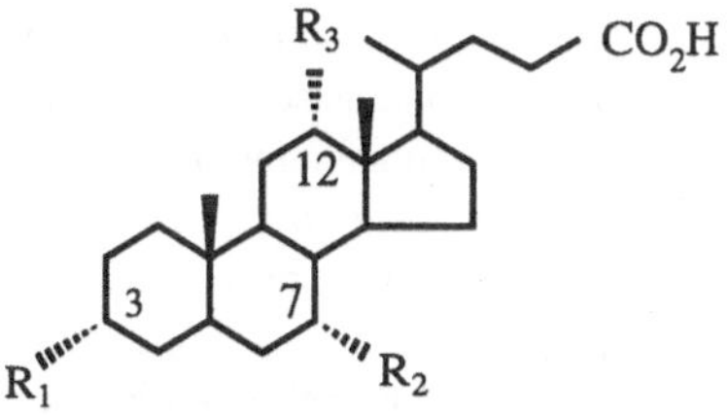

Figure 2 Steroids; cholic acid ($R_1=R_2=R_3=OH$) deoxycholic acid ($R_1=R_3=OH, R_2=H$).

STRUCTURE OF THE UNIVERSE

Hierarchical Structure of Particles

In order to understand the concept of the molecular information and their expression, we should first survey the hierarchical structure of particles or materials in the universe. After the Big Bang, various particles, such as quarks, nuclei, atoms, molecules, molecular assemblies and so on, were born with decreasing temperature, meaning that each particle has different bonding energy. This structure induces the following question; which particles function as information carriers. Generally speaking, the information originates from infinite kinds of combinations of simple elements. For example, computers utilize 0 and 1, and English does the alphabet. The combination of nuclei and electrons yielded only hundred kinds of atoms, while the combination of the atoms did infinite kinds of molecules. So, we reach the idea that organic molecules are the right information carriers.

M. Kamachi · A. Nakamura (Eds)
New Macromolecular Architecture and Functions
Proceedings of the OUMS '95 Toyonaka, Osaka, Japan, 2-5 June, 1995
© Springer-Verlag Berlin Heidelberg 1996

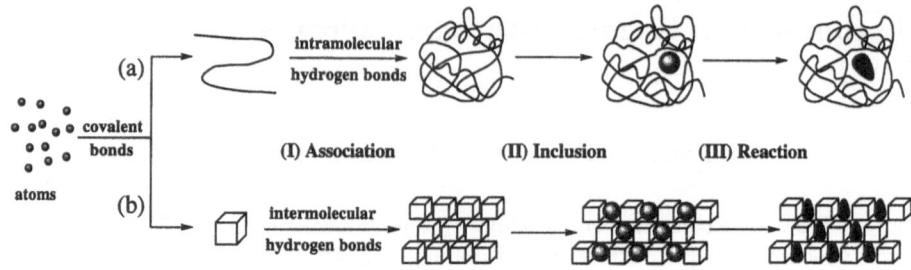

Figure 3   Fixation and expression of the molecular information
in the case of proteins (a) and assemblies (b).

Molecular Information and Their Expression

The next problem is how to express the information. Computers use pictures and sounds for this purpose, while the universe can use only physical forces. Among them, we can select weak interactions, such as hydrogen bonding and van der Waals forces, which work among the molecules. Careful consideration of the weak forces leads us to the following concept; the information can be fixed as the molecules by using strong covalent bonds, and expressed through the molecular assemblies by using weak noncovalent bonds.

This concept takes form in biological systems on earth. Naturally, proteins provide a good example. As shown schematically in Figure 3(a), a combination of atoms produces a sequential poly(α-amino acid) to fix the information. The polymeric chain folds mainly by intramolecular hydrogen bonds to yield a three-dimensional molecular architecture, which can include another organic substances. In some cases the included guests undergo any reactions. In this way we see that the expression consists of three steps; association, inclusion, and reaction.

Now, what about some organic molecules other than the biopolymers ? Since we have not hitherto discussed it, we do not have a clear definition in the case of these molecules. A criterion based on proteins should be whether the molecules exhibit the three steps. In addition, one thing we may ask is whether it is indispensable for the molecular architectures to exist in solution or not. From the fact that proteins work in aqueous solution, can we exclude crystals which small molecules or oligomers may form by intermolecular hydrogen bonding forces ? I believe that the solution is not essential, but the utilization of weak noncovalent interactions is essential for the expression of the molecular information instead.

Information Carriers Other Than Biopolymers

What kind of organic molecules are significant as the information carriers ? We have to consider some requisites for the carriers. For example, moderate monodispersed molecular weights, regulation of free rotation, multiple asymmetric carbons and heteroatoms and so on. Chemists can draw many kinds of linear chains, such as polymethylenes, polyoxyethylenes, polyiminoethylenes and so on. However, they are not free to synthesize the chains which satisfy the requisites mentioned above. This is primary because it is very difficult to introduce many asymmetric carbons and hydrogen bonding groups into the chains freely.

Instead, there are natural organic compounds having many asymmetric carbons and heteroatoms. For example, steroids, alkaloids, polysaccharides and so on are well-known. They have been so far regarded as important from a viewpoint of physiological activities, but now should be done from a viewpoint of the information carriers. Among them we focus on the steroids, which form crystalline assemblies through intermolecular hydrogen bonds. Many assemblies can include organic guests which undergo reations in some cases (Figure 3(b)).

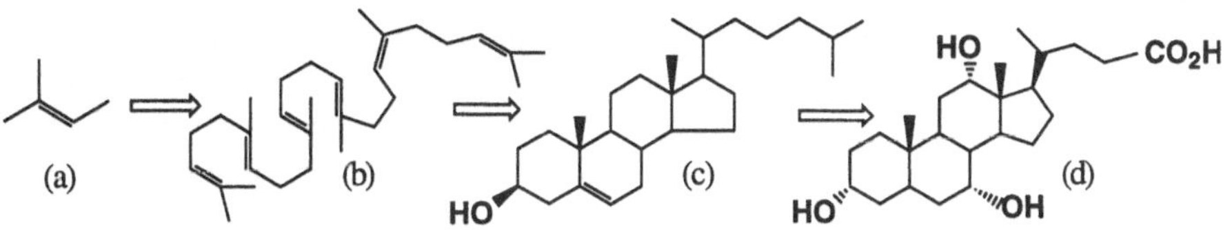

Figure 4   Molecular evolution of steroids from isoprene (a) through squalene (b) and cholesterol (c) to cholic acid (d).

STEROIDS

Bile Acids and Their Derivatives

The following consideration based on molecular evolution makes clear a structural significance of the steroids (Figure 4). The starting compound, isoprene, polymerizes to form a hexamer, called squalene. This linear carbon-chain compound undergoes cyclization followed by addition of hydrogen bonding groups to produce cholesterol. Further addition of the groups yields cholic acid. This indicates that the steroids are oligomeric carbon-chain compounds or copolymers of substituted methylenes.

Among so many steroidal compounds, we take notice of bile acids (Figure 5), particularly cholic(**a1**), deoxycholic(**b1**), chenodeoxycholic(**c1**) and lithocholic(**d1**) acids from two reasons. The one is that these acids involve multiple hydrogen bonding groups which facilitate weak noncovalent bonds. The other is that we can easily synthesize so many derivatives starting from the acids.

Sequential Carbon-Chain Oligomers

The secret of proteins lies in the sequence of peptide chains. This sequence is performed also for the carbon-chain compounds, since the steroidal acids can diversely be changed in the following way (Figure 5). First, carboxylic groups of side chains can be exchanged to amide,

a (R_1,R_2,R_3 / OH, OH, OH)	b (R_1,R_2,R_3 / OH, H, OH)	R_4
Cholic acid (**a1**)	Deoxycholic acid (**b1**)	1. CH_2CO_2H
Cholanamide (**a2**)	Deoxycholanamide (**b2**)	2. CH_2CONH_2
5β-Petromyzonol (**a3**)	3α,12α,24-Trihydroxy-5β-cholane (**b3**)	3. CH_2CH_2OH
Methyl cholate (**a4**)	Methyl deoxycholate (**b4**)	4. $CH_2CO_2CH_3$
N-Methylcholanamide (**a5**)	N-Methyldeoxycholanamide (**b5**)	5. $CH_2CONHCH_3$
Glycocholic acid (**a6**)	Glycodeoxycholic acid (**b6**)	6. $CH_2CONHCH_2CO_2H$
Taurocholic acid (**a7**)	Taurodeoxycholic acid (**b7**)	7. $CH_2CONHCH_2CH_2SO_3H$
Norcholic acid (**a8**)	Nordeoxycholic acid (**b8**)	8. CO_2H
Homocholic acid (**a9**)	Homodeoxycholic acid (**b9**)	9. $CH_2CH_2CO_2H$

c (R_1,R_2,R_3 / OH, OH, H)	d (R_1,R_2,R_3 / OH, H, H)	
Chenodeoxycholic acid (**c1**)	Lithocholic acid (**d1**)	
Chenodeoxycholanamide (**c2**)	Lithocholanamide (**d2**)	
3α,7α,24-Trihydroxy-5β-cholane (**c3**)	3α,24-Dihydroxy-5β-cholane (**d3**)	
Methyl chenodeoxycholate (**c4**)	Methyl lithocholate (**d4**)	
N-Methylchenodeoxycholanamide (**c5**)	N-Methyllithocholate (**d5**)	
Glycochenodeoxycholic acid (**c6**)	Glycolithocholic acid (**d6**)	
Taurochenodeoxycholic acid (**c7**)	Taurolithocholic acid (**d7**)	
Norchenodeoxycholic acid (**c8**)	Norlithocholic acid (**d8**)	
Homodeoxycholic acid (**c9**)	Homolithocholic acid (**d9**)	

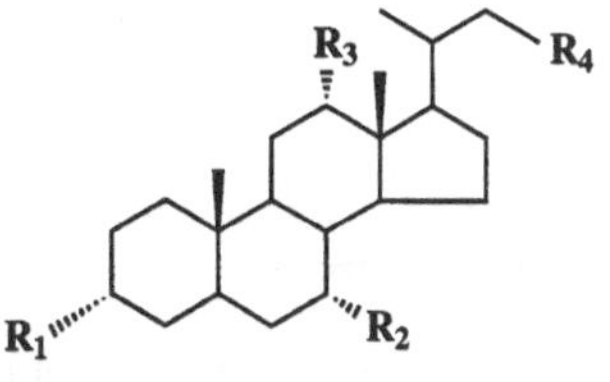

Figure 5   Steroidal derivatives and their abbreviation used in this article.

hydroxyl, ester groups and so on. Secondly, the carbon chains can be shortened or lengthened. Thirdly, the absolute configurations of hydroxyl groups of the skeletons are easily changed, leading to a preparation of over twenty kinds of the skeletons. These changes can be combined to yield over thousand derivatives, indicating that these steroids are sequential carbon-chain compounds.

MOLECULAR ARCHITECTURES OF THE STEROIDS

Cumulated Bilayered Assemblies

These steroids form crystalline molecular assemblies by recrystallizing from various solvents. We consider that analysis of the assemblies will enable us to exemplify the concept mentioned above. So, we determined their single-crystal structures by X-ray diffraction method, leading to detailed knowledge of intermolecular hydrogen bonds, molecular arrangements, host-guest interactions and so on.[3-12] We already analyzed over hundred crystals of the steroids since the discovery of inclusion abilities of **1a**.[3] It was found that most of them form layers, and few of them tubes. In the following section the layered assemblies will be discussed.

In order to understand the assemblies sterically, Figure 6 depicts a projection of an assembly

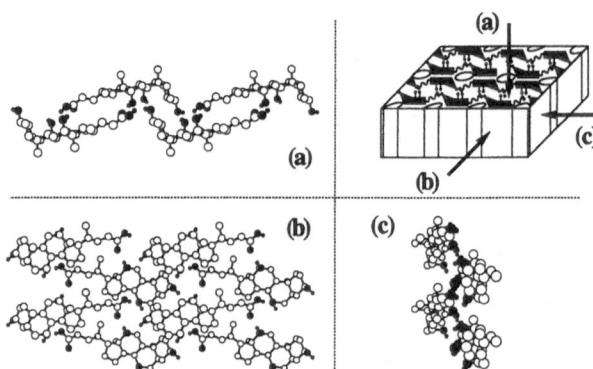

of steroidal molecules. This figure is composed of three different views; top, front and right-side views.

Figure 7 illustrates six kinds of the top views for the assemblies of their carboxylic, amide, alcoholic derivatives. It can be seen that they form characteristic bilayered crystals which are modified, depending on the chemical structures and guest components.

Figure 6 Projection of an assembly of steroidal molecules; (a) top view, (b) front view, (c) right-side view.

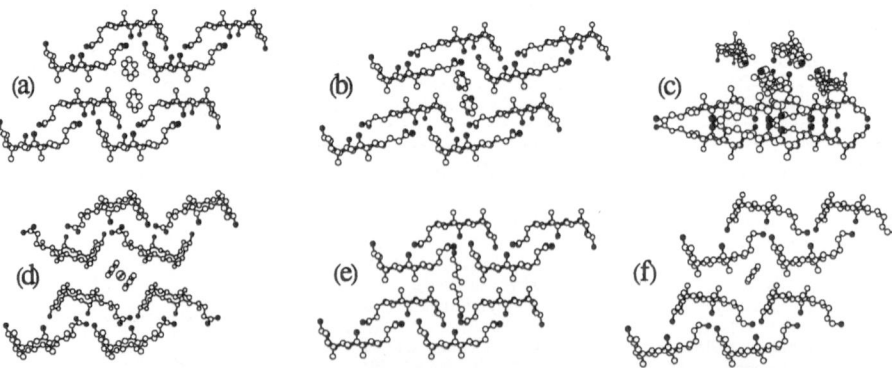

Figure 7 Top views of bilayered assemblies of bile acids and their derivatives in the inclusion crystals. The hosts; (a) cholic acid(**a1**), (b) cholanamide(**a2**); (c) 5β-petromyzonol(**a3**); (d) deoxycholic acid(**b1**); (e) deoxycholanamide(**b2**); (f) 3α,12α,24-trihydroxy-5β-cholane(**b3**).

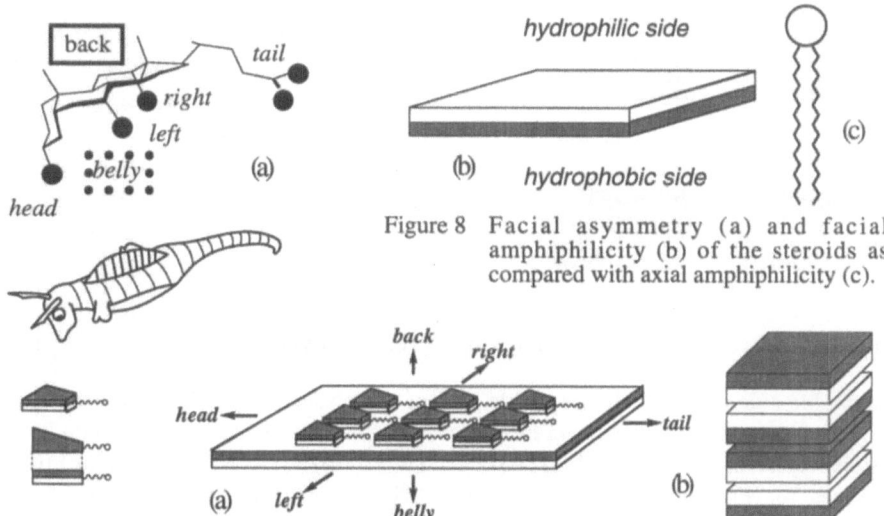

Figure 8 Facial asymmetry (a) and facial amphiphilicity (b) of the steroids as compared with axial amphiphilicity (c).

Figure 9 Formation of an amphiphilic and asymmetric sheet (a) and the stacking of the sheets (b).

Interpretation of the Architectures Based on Molecular Structures

Facial amphiphilicity and asymmetry : Bile acids and their derivatives have several conflicting structural elements, such as large and small, rigid and flexible, axial and facial, flat and wavy, symmetric and asymmetric, as well as polar and nonpolar parts. The skeleton is large, rigid and facial, while the side chain is small, flexible and axial. Such conflicts come from characteristic facial structures of the steroidal compounds as compared with axial ones (Figure 8). When the skeletons have multiple hydrophilic groups on the one side, the molecules will acquire the facial amphiphilicity. A combination of the asymmetry with the facial amphiphilicity and conflicts yields an example of clear facial asymmetry; three axes of the host molecules can be distinguished due to the asymmetry, just like vertebrate animals, above all a sea horse. Thus, we use the facial amphiphilicity of the steroidal skeleton. We call the hydrophilic side *belly* and the lipophilic side *back*. Second, we distinguish another direction by the large skeleton and the small side chain. We call the former part *head* and the latter *tail*. Third, the rest is based on the hydroxyl groups on the belly side. The hydroxyl groups at 12 and 7 positions operate as marks discriminating between *right* and *left*, respectively.

Molecular arrangements : The most simple arrangement of such asymmetric molecules is shown in Figure 9(a). The molecules arrange in a *head-to-tail* and *right-to-left* fashion, leading to a preparation of an amphiphilic and asymmetric sheet; one side of the sheet is hydrophilic and other is hydrophobic. For example, **1a** and their derivatives employ this arrangement in many cases, since they are facially amphiphilic.[3-6] **2a** and their derivatives usually adopt this arrangement, but in some cases they adopt another one.[7-12] On the other hand, **3a**, **4a** and their derivatives lose the facial amphiphilicity in many cases. So, they have different arrangements, particularly in a *right-and-left* direction, leading to disappearance of a *belly-and-back* direction.

Stacking of the sheets : The amphiphilic sheets stack together to form bilayers (Figure 9(b)). That is, the hydrophilic side of one sheet meets the same side of other sheet, while the

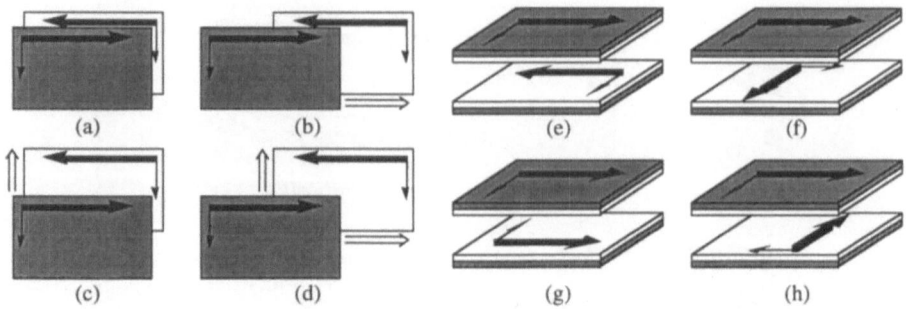

Figure 10 Slidings (a)-(d) and rotations (e)-(h) between the asymmetric and amphiphilic sheets.

hydrophobic side meets the hydrophobic side of other sheet. Since the sheets are highly asymmetric, there are many stacking methods. Figure 10 schematically shows eight kinds of plausible ways. The sheets can slide in two directions ((a)-(d)) and rotate each other ((e)-(h)). As can be seen from Figures 7(a) and (c), **a1** adopts a sliding mode (d) and a rotation mode (e),[3] while **b1** does the same (d) and a different (g).[7] **a2** and **b2** employ similar modes of the original carboxylic hosts in most cases.[4,8] In contrast, **a3** takes a rotation mode (f),[5] while **b3** takes (g).[9]

Hydrogen bonding networks : Three or four hydrogen bonding groups can combine together in many ways. Although there are principally 4x3x2x1=24 ways, we can observe parts of them. The most typical combination is seen in Figure 11. Four different host molecules provide one of each group, resulting in a cyclic network. The other is a helical network. Figure 12 shows the networks observed in the case of **a1**, **b1** and their derivatives. **a1** and **a2** form similar cyclic networks with retention of assembly modes of the host molecules.[3,4] Accordingly, this pair of the networks serve as a good example of evaluation of hydrogen bonds between the host and guest components. **a3** form a completely different network from **a1**,[5] while **b3** forms a similar one to **b1**.[9] The guests having hydrogen bonding groups are inserted into the networks, leading to a slight change of them.

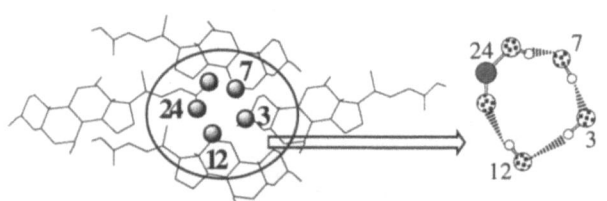

Figure 11 Unit of hydrogen bonding network composed of four host molecules.

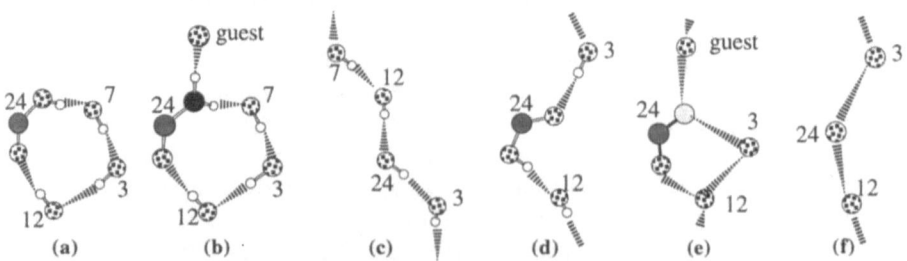

Figure 12 Various hydrogen bonding networks. (a)**a1**, (b)**a2**, (c)**a3**, (d)**b1**, (e)**b2** and (f)**b3**.

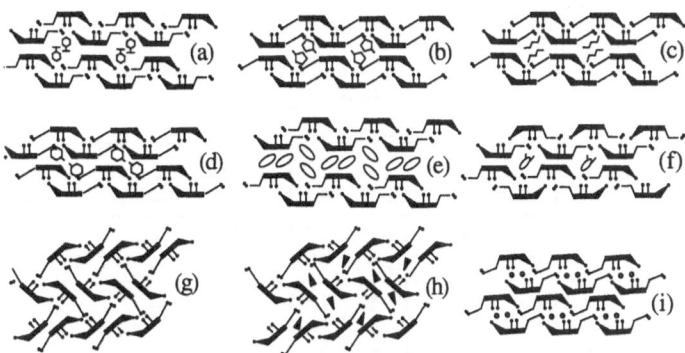

Figure 13
Polymorphic crystals
of **a1**. Guests;
(a) acetophenone
(b) γ-valerolactone
(c) acetylacetone
(d) 3-methyl-
cyclohexanone
(e) indene
(f) o-xylene
(g) none
(h) acrylonitrile
(i) water.

POLYMORPHIC CRYSTALS AND THEIR INTERCONVERSION

Polymorphic Crystals

Bile acids and their derivatives form guest-dependent polymorphic crystals owing to their multifunctionality and asymmetry. For example, Figure 13 shows a schematic representation of the guest-dependent polymorphism of **a1**.[1,3] In most of organic guests the host molecules form bilayers through cyclic hydrogen bonds to yield monoclinic crystals having versatile channels. The side chains adopt *gauche* (Figures 13(a)(e)(f)) and *trans* (Figures 13(b)(c)(d)) conformations, depending on the guests. Indene molecules insert into the lipophilic sides of the bilayers in addition to usual channels to yield a sandwich-type structure (Figure 13(e)). In another monoclinic crystal (Figure 13(f)), the host molecules arrange in a reverse way from those mentioned. The others are orthorhombic crystals ($P2_12_12_1$) having small cavities (Figures 13(g)(h)). Water forces the host to arrange in a slightly different way (Figure 13(i)).

Interconversion Based on Intercalation

Such polymorphic phenomena induce the following question; is it possible to intercovert these crystals reversibly ? We found that guest components can be replaced, inserted and eliminated briefly, when the host **a1** are sparingly soluble in liquid guests. So, we proposed the concept of intercalation in organic crystals for the first time.[13] One thing we pay attention is whether the architectures can be maintained during the intercalation or not. This is necessary for a comparison between proteins and organic crystalline assemblies. Figure 14 schematically shows the intercalation in the case of **a1**. The bilayered crystals have the hydrophobic sides which can easily slide during the replacement of guests, accompanied by the change of the conformations of the side chain and the hydrogen bonding networks. On the other hand, the insertion and elimination of a guest bring about a great change of the hydrogen bonding networks, leading to a collapse of the crystals in some cases. These results indicate that it takes place a dynamical change in the crystals.

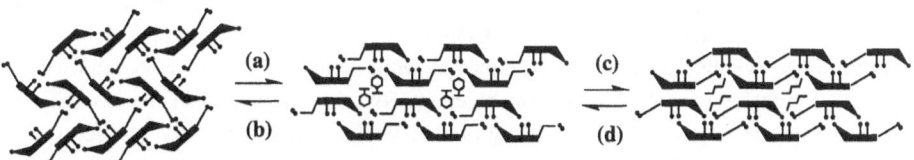

Figure 14 Interconversion of the polymorphic crystals of **a1** through insertion (a),
elimination (b) and replacement (c)(d) of guest components.

Table 1 Selective inclusion of aliphatic alcohols by using crystalline assemblies of steroidal hosts.

Guest		a1	b1	a2	b2	a3	b3
				Host			
C1	Methanol	1 : 1	- [b)]	1 : 1	1 : 1	GF	-
C2	Ethanol	1 : 1	-	1 : 1	1 : 1	GF	-
C3	1-Propanol	1 : 1	-	1 : 1	1 : 1	GF	-
	2-Propanol	GF [a)]	1 : 1	1 : 1	1 : 1	GF	-
C4	1-Butanol	GF	-	1 : 1	1 : 1	GF	-
	2-Butanol	GF	-	1 : 1	1 : 1	GF	-
	2-Methyl-1-propanol	GF	2 : 1	1 : 1	1 : 1	GF	-
	2-Methyl-2-propanol	GF	2 : 1	1 : 1	1 : 1	GF	-
C5	1-Pentanol	GF	5 : 2	1 : 1	1 : 1	GF	-
	2-Pentanol	GF	2 : 1	1 : 1	1 : 1	GF	-
	3-Pentanol	GF	5 : 2	1 : 1	1 : 1	GF	-
	2-Methyl-1-butanol	GF	2 : 1	1 : 1	1 : 1	GF	-
	2-Methyl-2-butanol	GF	2 : 1	1 : 1	1 : 1	GF	-
	3-Methyl-2-butanol	GF	2 : 1	1 : 1	1 : 1	GF	-

a) GF : Guest-Free Crystal   b) - : Not Crystallized

INCLUSION AND RECOGNITION

Since the discovery of inclusion abilities of **a1**, we already confirmed the abilities of over twenty kinds of bile acid derivatives. Interestingly, except for a few cases they included any organic guests. Some hosts include a wide variety of organic substances, while other hosts include a few guests. For example, **b5** includes only one kind of guest.[11] Accordingly, this inclusion behaviour serves as an efficient example for molecular recognition in solid states. Each architecture has the corresponding cavity, depending on the association modes and the hydrogen bonding networks.

We found an example for showing an role of the hydrogen bonding networks. Table 1 shows the inclusion of aliphatic alcohols by various steroidal hosts. **a1** includes only three small alcohols, while **b1** does not include them. On the other hand, amide

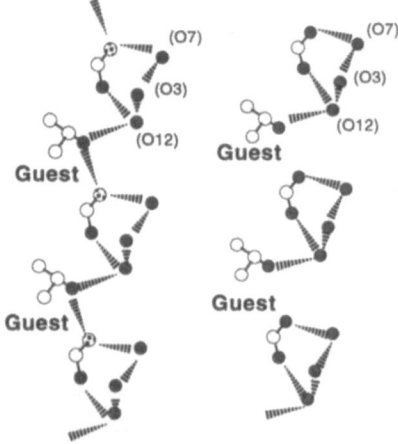

Figure 15
Stable inclusion of alcoholic guests by double hooks (a) as compared with that by a plausible mono hook (b).

hosts, **a2** and **b2**, include all the alcohols employed, while **a3** and **b3** do not . How can we explain this result ? Consideration of the hydrogen bonding networks gave us a plausible mechanism for that. The alcoholic guests are caught by double hooks between the two cyclic networks composed of the host molecules, as shown in Figure 15. But the acid hosts probably form a mono hook. An extra cavity may permit some guests to move there actively, resulting in a collapse of the inclusion state.

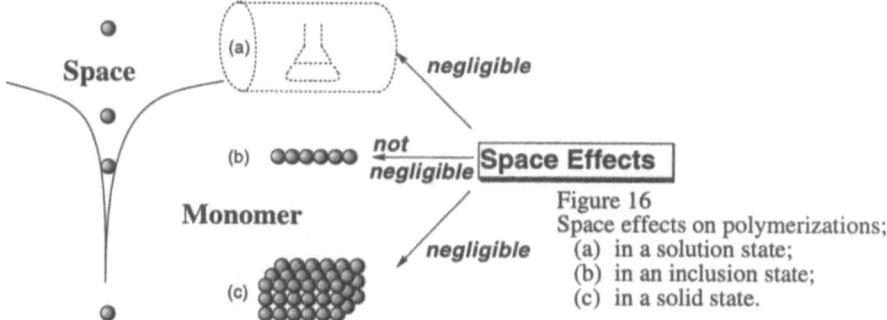

Figure 16
Space effects on polymerizations;
(a) in a solution state;
(b) in an inclusion state;
(c) in a solid state.

POLYMERIZATION REACTION

We studied on inclusion polymerization by using a pair of hosts, **b1** and apocholic acid, in detail.[14] The polymerization is based on one-dimensional assemblies of monomers in molecular channels. The study caused the simple idea that space effects can be efficiently obseved for the inclusion polymerization, whereas the effects are negligible in the case of solution and solid polymerizations. Figure 16 schematically shows such a situation. In addition, the idea reminds us of a daily relation between the spaces and included guests, such as a person in a room, a notebook in a bag and so on. Namely, the relation is observed on both a macroscopic and microscopic level, indicating that the relativity between the host and guest is the most important factor.

OVERVIEW OF OUR RESEARCH

Figure 17 shows an overview of our research performed by using steroidal hosts. Starting from the inclusion polymerization, we extended the study to various host-guest relationships. The research directed me to establish the concept; the molecular information and their expression. This development comes from the fact that there exists a substantial similarity between proteins and the steroidal assemblies in spite of a great difference of their molecular weights.

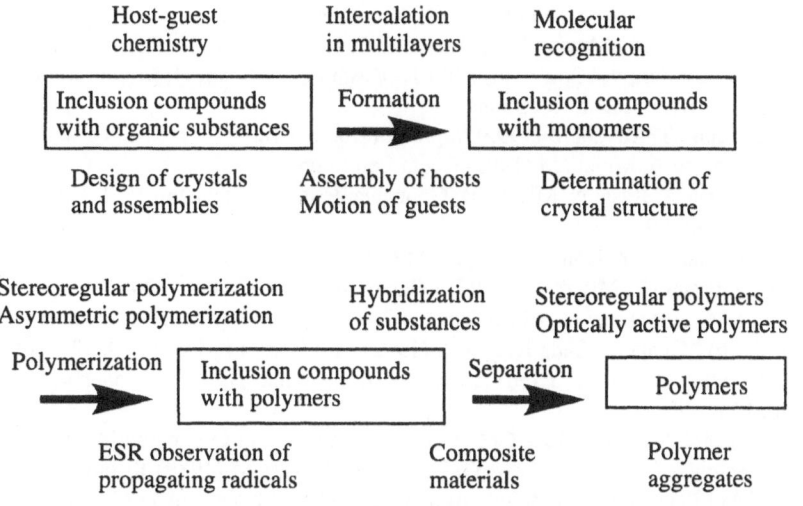

Figure 17   Overview of our research performed by using steroidal inclusion compounds.

30

CONCLUSION

Although vinyl polymers such as polyethylene have been known for a long time, the issue of sequential carbon-chain polymers still remains unsolved. Accordingly, people use the concept of the molecular information only in the case of biopolymers. However, steroids tell us that there exist naturally-occurring known oligomers suitable for the information carriers. This reminds me of languages; we know poetry involving letters of the English alphabet or *Haiku* and *Tanka* of the Japanese characters (seventeen and thirty-one characters, respectively) as compared with novels of the very long ones. It is considered that even oligomeric combinations of various elements can create the significant information. Finally, I believe that chemistry becomes more attractive by introduction of the concept of the molecular information and their expression.

References
1   Miyata M, Sada K In: Lehn M (ed) Comprehensive supramolecular chemistry Vol.5, in press
2   a) Miyata M (1992) Biophysics 32:169 ; b) Miyata M (1993) Nippon Kagaku Kaishi J Chem Soc Jap 128 ; c) Miyata M (1994) Polymer Applications 43:560
3   a) Miyata M, Shibakami M, Goonewardena W, Takemoto K (1987) Chem Lett 605 ; b) Miki K, Masui A, Kasai N, Miyata M, Shibakami M, Takemoto K (1988) J Amer Chem Soc 110: 6594 ; c) Miyata M, Shibakami M, Takemoto K (1988) J Chem Soc Chem Commun 655 ; d) Miki K, Kasai N, Shibakami M, Chirachanchai S, Takemot k, Miyata M (1990) Acta Cryst C46:2442 ; e) Miki K, Kasai N, Shibakami M,Takemoto K, Miyata M (1991) J Chem Soc Chem Commun 1757 ; f) Miyata M,Sada K, Hirayama K, Yasuda Y, Miki K (1993) Supramol Chem 2:283 ; g) Nakano K, Sada K, Miyata M (1994) Chem Lett 137 ; h) Nakano K, Sada K, Miyata M (1994) J Chem Soc Chem Commun 953
4   a) Sada K, Kondo T, Miyata M, Tamada T, Miki K (1993) J Chem Soc Chem Commun 753 ; b) Sada K, Kondo T, Miyata M, Miki K (1994) Chem Mater 6:1103 ; c) Sada K, Kondo T, Miyata M (1995) Tetrahedron Asym in press
5   Sada K, Kondo T, Yasuda Y, Miyata M, Miki K (1994) Chem Lett 727
6   a) Miyata M, Goonewardena W, Shibakami M, Takemoto K, Masui A, Miki K, Kasai N (1987) J Chem Soc Chem Commun 1140 ; b) Miki K, Masui A, Kasai N, Goonewardena, W, Shibakami, M, Takemoto, Miki M (1992) Acta Cryst C48:503
7   a) Giglio E (1984) In: Atwood JL, Davies JED, NacNicol DD (eds) Inclusion compounds. Academic, London, vol 2, 207 ;b) Miki K, Kasai N, Tsutsumi H, Miyata M, Takemoto K (1987) J Chem Soc Chem Commun 545
8   Sada K, Kondo T, Miyata M (1995) Supramol Chem 5:189
9   Sada K, Matsuo A, Miyata M (1995) Chem Lett in press
10 Miki K, Masui A, Kasai N, Miyata M, Goonewardena W, Shibakami M, Takemoto K (1989) Acta Cryst C45:79
11 Sada K, Hishikawa Y, Kondo T, Miyata M (1994) Chem Lett 2113
12 Sada K, Kitamura T, Miyata M (1994) J Chem Soc Chem Commun 905
13 a) Miyata M, Shibakami M, Chirachanchai S, Takemoto K, Kasai N, Miki K (1990) Nature 343:446 ; b) Miyata M, Sada K, Hori S (1992) Mol Cryst Liq Cryst 219: 71 ; c) Miyata M, Miki K (1993) In: Ohashi Y (ed) Reactivity in molecular crystals. Kodansha-VCH, Tokyo Berlin, 153
14 a) Miyata M(1992) In: Paleos CM (ed) Polymerization in Organized Media. Gordon and Breach Science Publishers, New York, 327 ; b) Miyata M (1989) Kobunshi 38:1006 ; c) Miyata M (1991) Kobunshi 40:398; d) Miyata M (1992) Kobunshi 41:606 ; e) Miyata M In: Lehn M (ed) Comprehensive supramolecular chemistry Vol.10, in press

Stereospecific Living Polymerization of Methacrylate with Binary Initiator Systems

Tatsuki Kitayama and Koichi Hatada

Department of Chemistry, Faculty of Engineering Science,
Osaka University, Toyonaka, Osaka 560, Japan

Abstract: Three kinds of stereospecific polymerizations of methacrylate are described; isotactic-specific, syndiotactic-specific, and heterotactic-specific ones, all of which utilize binary initiator systems comprising bulky anionic initiator and Lewis acid; the latter works as stereospecificity modifier. Particular emphasis is placed on the unique synthetic utility of bulky aluminum phenoxides in methacrylate polymerization; heterotactic polymerization of primary and secondary alkyl esters, syndiotactic polymerization of tertiary esters, monomer-selective copolymerization leading to the formation of block copolymer with high stereoregularity, regioselective polymerization of a dimethacrylate, affording a soluble, reactive polymer, and polymerization of methacrylate having sterically protected phenol group by anionic mechanism.

INTRODUCTION

Structure control of polymer molecule is a primary step toward precise control of polymer properties that is required for the advanced polymeric materials such as functional and specialty polymers. Living polymerization, the concept of which was first proposed by Szwarc in 1956 [1], is one of the most promising ways for controlling the molecular weight and its distribution (MWD) as well as the structure of end groups. Another important structural feature to be controlled is stereoregularity. With one preceding finding of stereoregular poly(vinyl ether) by Schildknecht [2] in 1947, the field of stereospecific polymerization actually came into existence when Ziegler [3] and Natta [4] developed new polymerization systems which exhibited unique stereoregulating powers in olefin polymerization. For the precise control of structures of polymer molecule, combination of these two types of polymerizations is desirable, that is, stereospecific living polymerization. The term "stereospecific living polymerization" was first proposed by Soum and Fontanille in 1980 [5] regarding an isotactic (*it*-) living polymerization of 2-vinylpyridine in benzene with an organomagnesium compound. Later a syndiotactic (*st*-) living polymerization was also reported for 2-isopropenylpyridine in tetrahydrofuran (THF) with alkyllithium [6].

Methacrylate is one of the most extensively studied classes of vinyl monomer, particularly, in regard of stereoregularity of the obtained polymers. The stereoregularity is a function of

M. Kamachi · A. Nakamura (Eds)
New Macromolecular Architecture and Functions
Proceedings of the OUMS '95 Toyonaka, Osaka, Japan, 2-5 June, 1995
© Springer-Verlag Berlin Heidelberg 1996

monomer structure (structure of ester group), initiator, solvent, temperature and so on. Proper selection of conditions allows preparation of a wide variety of stereoregular polymethacrylates [7,8]. On the other hand, living polymerizations of the polar monomers have been developed in the last decade [9]. However, living polymerizations which afford stereoregular polymers have still been limited. Yasuda and his coworkers found that organolanthanide complexes such as $[(C_5Me_5)_2SmH]_2$ give high molecular weight st-poly(methyl methacrylate)s (PMMAs) with narrow MWD in a living manner [10].

In the 1980's, we found two stereospecific living polymerizations of methacrylate; one is it-specific, initiated with t-butylmagnesium bromide (t-BuMgBr) [11,12], and the other st-specific, initiated with t-butyllithium/trialkylaluminum (t-BuLi/R$_3$Al) [13,14]. More recently, we also found heterotactic-(ht-) specific living polymerization of certain alkyl methacrylates with a combination of t-BuLi and bis(2,6-di-t-butylphenoxy)methylaluminum [MeAl(ODBP)$_2$] (Al/Li =5) in toluene at low temperature [15-18]. These stereoregular polymethacrylates have the same chemical structure; t-butyl group at one end and a methine hydrogen at the other end, and are very appropriate samples for testing tacticity dependencies of the properties of the polymers. In this paper are described these stereospecific living polymerizations of methacrylate with a particular emphasis on the latest one, t-BuLi/MeAl(ODBP)$_2$, and its exploitation to the control of methacrylate polymerization and copolymerization.

RESULTS AND DISCUSSION

Isotactic and Syndiotactic Living Polymerizations

Isotactic PMMA is usually prepared by anionic initiators such as alkyllithium and Grignard reagent in non-polar solvents. However, most of the polymerization reactions cannot be fully controlled owing to side reactions and multiplicity of active species which make the MWD of the polymer broad [8]. It has been revealed for some polymerization systems that the addition of the initiator to the C=O bond of methyl methacrylate (MMA) in the initial stage of polymerization is the major side reaction that causes the complexity in the proceeding reaction process.

To prevent such complexity, we utilized a sterically bulky Grignard reagent, t-BuMgBr, prepared in diethyl ether, as an initiator for the polymerization of MMA in toluene at low temperature [11]. The polymerization proceeds without side reactions owing to the bulkiness of t-butyl group of the initiator and gives a highly isotactic polymer with narrow MWD (Table 1) [11,12]. The ether solution of t-BuMgBr contains excess amount of MgBr$_2$ produced by an Wurtz-type coupling during the preparation, which contributes to the formation of the propagating species with homogeneous activity and high stereospecificity [12]. Thus the initiator may be better described as a binary system, t-BuMgBr/MgBr$_2$, though the exact structure is not clear.

Highly syndiotactic living polymerization of MMA was realized by use of combinations of t-BuLi and R$_3$Al (Table 1) [13,14]. The polymerization by t-BuLi alone gives an isotactic-rich polymer with broad MWD, and the Mn of the polymer is much larger than the calculated value. Addition of R$_3$Al such as (n-Bu)$_3$Al decreases the isotacticity accompanied by an increase in syndiotacticity. With an increase in Al/Li ratio the syndiotacticity increases and the MWD becomes narrower, and highly syndiotactic PMMAs with narrow MWD are formed at the ratios of Al/Li$\geqq$2. The Mn values were close to the values calculated from the amounts of t-BuLi and the monomer consumed [13,14]. t-BuLi and R$_3$Al does not form an ate complex as far as they are mixed at the polymerization temperature (-78°C) as evidenced by ^{1}H NMR analysis of the mixture [14]. The st-polymethacrylates obtained contain t-butyl group at the initiating chain-end and a methine hydrogen at the terminating chain-end, but no alkyl groups originated form the aluminum component. The end-group structure indicates the initiation from t-BuLi and rules out the possibility of initiation from an ate complex such as t-BuR$_3$Al$^-$Li$^+$. The aluminum component is assumed to coordinate to the propagating chain-end to stabilize it and to alter the stereospecificity from isotactic to syndiotactic, and also to coordinate to the monomer to activate it. The monomer activation by the aluminum compounds is assumed from the observation of the increase in polymer yield with an increase in Al/Li ratio [14], though the kinetics was found not to obey a simple "monomer-activation mechanism" [19].

Heterotactic Living Polymerization

Heterotactic polymer is one type of stereoregular polymer that comprises an alternating sequence of meso (m) and racemo (r) diads. The formation of ht-sequence requires two different types of stereoregulation, m and r additions, to occur in an alternate manner, and thus

Table 1   Stereospecific living polymerization of methacrylate in toluene at -78 or -95°C[a]

Initiator	Ester	Temp. (°C)	Yield (%)	$\overline{Mn}$	$\dfrac{\overline{Mw}}{\overline{Mn}}$	Tacticity(%) mm	mr	rr	Ref.
t-BuMgBr[b]	Methyl	-78	73	3560	1.14	96.3	3.6	0.1	12
t-BuLi/n-Bu$_3$Al(1/3)	Methyl	-78	100	5510	1.17	0	8	92	14
t-BuLi/MeAl(ODBP)$_2$(1/5)	Methyl	-78	100	11640	1.14	11.6	67.8	20.6	15
	Ethyl	-78	100	7010	1.08	7.6	87.2	5.2	16
	Ethyl[c]	-95	100	8100	1.11	7.1	91.6	1.3	17
	n-Butyl	-78	98	9300	1.07	8.4	87.1	4.5	16
	t-Butyl[c]	-78	19	3050	1.19	7.4	8.5	84.1	16
	Me$_3$Si	-78	100	7030	1.16	0.3	3.3	96.4	29
	Me$_3$Si[c]	-95	48	3290	1.04	0.5	1.4	98.1	29

a Monomer 10mmol, initiator 0.2mmol, toluene 10ml, polymerization time 24h.

b [Mg]/[t-Bu$^-$] = 2.2.   c Polymerization time 48h.

the propagating species, an *st*-polymer forms in low yields. The result suggests that the propagating species are stabilized by the coordination with the bulky aluminum phenoxide, becomes less reactive, and favors *r* addition with the monomer free from the coordination by the aluminum phenoxide. In the polymerization system at higher Al/Li ratios, the excess MeAl(ODBP)$_2$ may activate the monomer through coordination, and the less reactive propagating species preferentially add the activated monomer, as shown in Scheme 1. The steric interaction between the sterically crowded active-end and the bulky monomer-MeAl(ODBP)$_2$ complex might be an important factor for the *ht*-propagation.

Stereochemical sequence distribution in *ht*-poly(EMA), estimated from ^{13}C NMR spectrum, is well characterized by first-order *Markovian* statistics; the conditional probabilities at -95°C, P*r*/*m*=0.972, P*m*/*r*=0.866; at -78°C, P*r*/*m*=0.926, P*m*/*r*=0.833, where for example P*r*/*m* is the probability of adding an *m* diad to an *r* chain end. The larger P*r*/*m* than P*m*/*r* means that $\sim\sim r\,M^-$ ended anion favors *m* addition with higher selectivity than $\sim\sim m\,M^-$ ended anion favors *r* addition. ^{13}C NMR analysis for the stereochemical triad at the terminating chain-end revealed that living polymer having *r* diad at the chain end exists more than that with *m* diad, suggesting the higher stability of the *r*-ended anion. The results seem to be consistent with the higher selectivity of the *r*-ended anion to form *ht*-sequence as observed from the larger P*r*/*m* [17].

The ^{13}C NMR analysis for the stereoregularity at and near the initiating chain-end revealed the stereospecificity of the dimer and trimer anions. The dimer anion preferentially undergoes *r* addition to form a trimer anion with *r* diad (*t*-Bu-M-$M^- \longrightarrow$ *t*-Bu-*r*-M^-). The trimer anion with *r* diad favors *m* addition (*t*-Bu-*r*-$M^- \longrightarrow$ *t*-Bu-*r*-*m*-M^-), and the *m*-selectivity is much enhanced by lowering the polymerization temperature as in the case of $\sim\sim r\,M^-$ ended anion [17].

Binary Initiator Systems for Stereospecific Living Polymerization

Eventually, most of the living polymerization systems explored in the 1980's utilize binary initiators. These can be categorized in three types of combination; (1) initiator + stabilizer , (2) initiator of low initiating ability + activator, and (3) initiator without initiating ability by itself + activator. Case (1) includes alkyllithium/LiCl system developed by Teyssie's group [20], *t*-BuLi with dialkylaluminum phenoxides such as (2,6-di-*t*-butylphenoxy)diisobutylaluminum by Ballard *et al.* [21]. Case (2) includes aluminum porphyrin complex in combination with bulky aluminum phenoxide [22]. Case (3) includes ketene silyl acetal and Lewis acid or base catalyst, so-called group transfer polymerization [23], tertiary phosphine/R$_3$Al [24], and enamine/ aluminum phenoxide system [25].

The *it*-specific initiator, *t*-BuMgBr, consists of "*t*-BuMgBr" and "MgBr$_2$". Thus, all initiators for the three stereospecific living polymerizations described in the preceeding sections are binary initiator systems comprising bulky anionic initiator, *t*-BuMgBr or *t*-BuLi, and the second component which does not initiate polymerization but affects the stereoregulation.

the diad configuration at the propagating chain-end should affect the stereospecificity drastically; the chain end with m diad prefers r-addition and the chain end with r diad prefers m-addition:

$$\sim\sim\sim m M^* \longrightarrow \sim\sim\sim m r M^*$$
$$\sim\sim\sim r M^* \longrightarrow \sim\sim\sim r m M^*$$

(M^*; last monomeric unit)

This obviously requires a higher-order stereoregulation than those for it- and st- polymerizations; for the latter the control of configurational relationship between neighboring constitutional repeating units is sufficient in principle.

Recently, we found that a combination of t-BuLi and MeAl(ODBP)$_2$ is an ht-specific initiator for the living polymerization of several alkyl methacrylates in toluene at low temperature. The polymerization of MMA with t-BuLi/MeAl(ODBP)$_2$ (Al/Li=5) gives heterotactic-rich PMMA with narrow MWD [15] (Table 1). Other alkyl methacrylates also give ht-polymers under the same conditions; in particular, ethyl (EMA) and butyl methacrylates give ht-polymers with mr triad of 87% [16] (Table 1). Although the initiator efficiencies are not 100%, the polymerization shows a living character. The highest heterotacticity so far attained is 91.6% in triad for poly(EMA) obtained at -95°C. Figure 1 shows the carbonyl carbon NMR signals for three types of stereoregular poly(EMA)s obtained by t-BuMgBr, t-BuLi/n-Bu$_3$Al, and t-BuLi/MeAl(ODBP)$_2$.

The existence of an excess amount of MeAl(ODBP)$_2$ over t-BuLi is essential for the formation of the ht-polymer. In fact, at the ratio of Al/Li=1, where most of the aluminum phenoxide might coordinate with

Fig. 1 125MHz ^{13}C NMR spectra of (a)isotactic, (b)syndiotactic, and (c)heterotactic poly(EMA)s (nitrobenzene-d$_5$, 110°C)

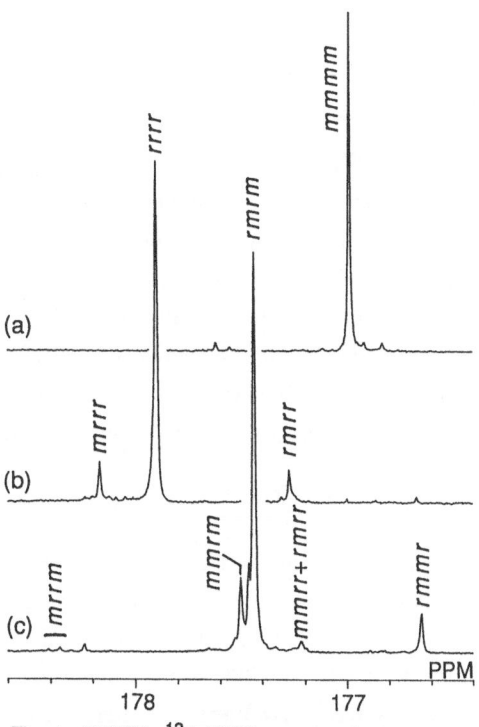

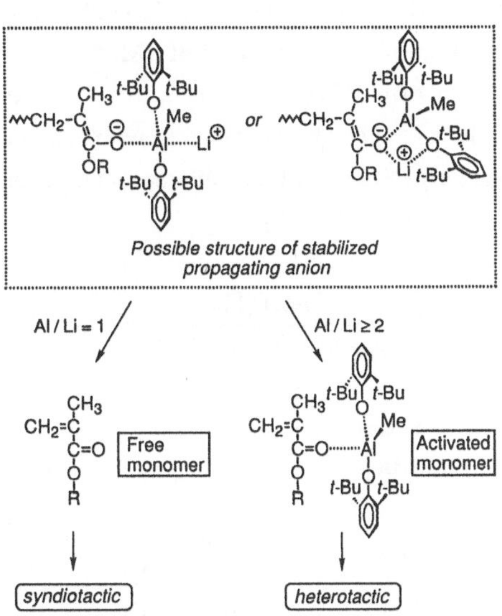

Possible structure of stabilized propagating anion

Scheme 1 Possible mechanism of the polymerization with t-BuLi/MeAl(ODBP)$_2$

36

CH₃ CH₃ CH₃ CH₃
—CH₂—C—CH₂—C—CH₂—CH₂—C—
C=O C=O C=O C=O
OR OR OR OR
isotactic
[*t*-BuMgBr / MgBr₂]

CH₃ C=O CH₃ C=O
—CH₂—C—CH₂—C—CH₂—C—CH₂—C—
C=O CH₃ C=O CH₃
OR OR
syndiotactic
[*t*-BuLi / R'₃Al]

OR OR
CH₃ CH₃ C=O C=O
—CH₂—C—CH₂—C—CH₂—C—CH₂—C—
C=O C=O CH₃ CH₃
OR OR
heterotactic
[*t*-BuLi / MeAl(ODBP)₂]

When tertiary phosphine is used instead of *t*-BuLi in combination with R₃Al and MeAl(ODBP)₂, *st*- and *ht*-polymers, respectively, of alkyl methacrylates are produced. Thus the main factor that controls the stereospecificity is owing to the second component instead of the initiator itself. These results imply the possibility that one can manipulate the stereospecificity of methacrylate polymerization by the combination of initiator and stereospecific modifier; the initiator by itself does not necessarily have stereoregulating powers and the modifier component which by itself does not initiate polymerization but provides stereospecificity.

Syndiotactic Polymerization of Trimethylsilyl Methacrylate with *t*-BuLi/MeAl(ODBP)₂

The stereospecificity of *t*-BuLi/MeAl(ODBP)₂ strongly depends on the structure of the ester group, and, interestingly, *t*-butyl (*t*-BuMA) and trimethylsilyl (TMSMA) methacrylates give *st*-polymers (Table 1). In particular, polymerization of TMSMA with *t*-BuLi/MeAl(ODBP)₂ (1/5) in toluene at -78°C gives *st*-poly(methacrylic acid) [poly(MAD)] (*rr* triad 96%) with narrow MWD (Mw/Mn=1.16) after quenching the reaction with methanol [26]. The polymerization at -95°C gives an *st*-polymer with *rr* triad of 98%, which is the highest value so far reported for polymethacrylate. In sharp contrast, polymerization with *t*-BuLi alone in toluene at -78°C gives an *it*-polymer (*mm*=95%) with fairly narrow MWD (Mw/Mn=1.31). Thus, when MeAl(ODBP)₂ is added to the polymer anion formed with *t*-BuLi, followed by the addition of the second feed of TMSMA, the polymerization proceeds further to give a polymer with narrow MWD (Mw/Mn=1.20). As expected, the polymer obtained comprises of *it*- and *st*-poly(MAD) blocks, so-called stereoblock polymer (*mm* : *mr* : *rr* = 44.2 : 5.8 : 50.0). By taking the tacticity of the first block into account, the tacticity of the second block is estimated as *mm* : *mr* : *rr* = 1.2 : 8.2 : 90.7 [27].

CH₃
CH₂=C
C=O
O
Si(CH₃)₃
TMSMA

t-C₄H₉Li
toluene/-78°C
→

CH₃ CH₃ CH₃
t-C₄H₉—(CH₂—C—CH₂—C)ₚ—CH₂—C⊖
C=O C=O C=O
O O O
Si(CH₃)₃ Si(CH₃)₃ Si(CH₃)₃
it-poly(TMSMA) anion

1)MeAl(ODBP)₂
2)TMSMA
3)HCl/MeOH
→

CH₃ CH₃ OH
C=O CH₃
t-C₄H₉—(CH₂—C—CH₂—C)ₚ—(CH₂—C—CH₂—C)q—H
C=O C=O CH₃ C=O
OH OH OH
stereoblock poly(MAD)

As seen in Scheme 1, monomer activation with MeAl(ODBP)$_2$ is an important process in the polymerization with t-BuLi/MeAl(ODBP)$_2$. As a consequence, the steric bulkiness in the ester group affects not only stereospecificity but also reactivity. In fact, rates of polymerization of these bulky monomers are rather small, probably due to hindered coordination by MeAl(ODBP)$_2$, leading to less effective activation of the monomer. Copolymerization of TMSMA with EMA, a less bulky monomer, proceeds in a monomer-selective manner without loosing stereospecificity for the both monomers, giving a stereoregular block copolymer that comprises ht-poly(EMA) and st-poly(TMSMA) blocks; the latter block is easily converted to st-poly(MAD) block [28]. High regularity of the copolymer as compared with

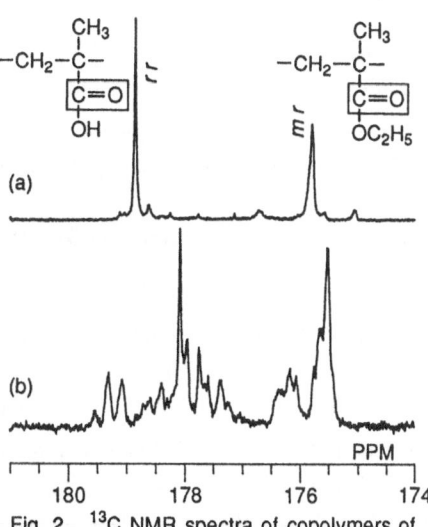

Fig. 2 ^{13}C NMR spectra of copolymers of MAD and EMA derived from copolymers of TMSMA and EMA prepared with (a) t-C$_4$H$_9$Li-MeAl(ODBP)$_2$ and (b) t-C$_4$H$_9$Li in toluene at $-78°C$ (125MHz , DMSO-d_6, 80°C)

that obtained by the copolymerization with t-BuLi alone is evident from their carbonyl carbon NMR signals as shown in Fig. 2.

The product formed at 50% conversion comprises almost exclusively EMA units, indicating that EMA polymerizes first preferentially, and then TMSMA is polymerized by the poly(EMA) anion to yield the block copolymer. However, ^{1}H NMR spectroscopic analysis for the end group (t-butyl group) reveals that in the initiation process t-BuLi adds both the monomers giving t-Bu-EMA and t-Bu-TMSMA sequences at the ratio of 2:3, i.e., k_i(EMA)/k_i(TMSMA)=2/3, k_i being the rate constant of initiation.

The results mean that t-BuLi itself scarcely shows monomer selectivity but the propagating methacrylate anion, stabilized through the coordination by MeAl(ODBP)$_2$, exhibits the high monomer selectivity. A similar monomer-selective polymerization is also possible for a combination of TMSMA and t-BuMA, where TMSMA polymerizes at -78°C preferentially and t-BuMA polymerizes after the temperature is raised to -40°C, giving an st-block copolymer with narrow MWD *(Mw/Mn=1.12)* . Terpolymerization of EMA, TMSMA and *t*-BuMA in one feed with *t*-BuLi-MeAl(ODBP)$_2$ in toluene at -78°C in the first stage and -40°C in the second stage affords a triblock copolymer comprising *ht*-poly(EMA), *st*-poly(MAD) and *st*-poly(*t*-BuMA) blocks. Thus the polymerization with *t*-BuLi/MeAl(ODBP)$_2$ displays the ability to control molecular weight, stereoregularity, and monomer sequence.

Regioselective Polymerization of Dimethacrylate with *t*-BuLi/MeAl(ODBP)$_2$

Based on the above findings of monomer selectivity by *t*-BuLi/MeAl(ODBP)$_2$, a dimethacrylate monomer [A] having primary and tertiary ester groups was synthesized and polymerized with *t*-BuLi-MeAl(ODBP)$_2$ (1/5) in toluene [32]. The polymerization at -60°C proceeds preferentially through the primary ester side and gives a soluble polymer having pendant methacrylate group along the chain. ^{1}H NMR analysis indicates that the polymer contains about 2 *t*-butyl group per chain, that is, about one branch per chain. Polymerization with *t*-BuLi itself resulted in the formation of an insoluble polymer. Thus MeAl(ODBP)$_2$ provides regioselectivity in this polymerization.

Anionic Polymerization of Methacrylate Having Sterically Protected Phenol Group

In the first report on the *ht*-polymerization of MMA [15], we mentioned the polymerization of MMA with *t*-BuLi in the presence of a five-fold excess of a reaction mixture of Me$_3$Al and 2,6-di-*t*-butylphenol in 1:3 molar ratio, which was found by ^{1}H NMR analysis to be a 1:1 mixture of MeAl(ODBP)$_2$ and the unreacted phenol. The fact that an *ht*-PMMA was obtained even in the presence of five-fold excess of the free phenol over the initiator, motivated our interest in polymerizing anionically a methacrylate monomer [B] with an unprotected hydroxy group having an identical steric hindrance as in the case of 2,6-di-*t*-butylphenol.

Polymerization of [B] with *t*-BuLi/MeAl(ODBP)$_2$ (1/5) in toluene at -78°C for 24 hr gives a polymer with narrow MWD *(Mw/Mn=1.08)* in 60% yield, whose tacticity was rich in heterotacticity; *mm* : *mr* : *rr* = 40 : 49 : 11. The initiator efficiency is less than unity (28%), indicating the occurrence of quenching of *t*-BuLi with the hydroxyl group of [B] in the initial stage of polymerization. The narrow MWD, however, suggests that once the initiation starts, a methacrylate anion formed, which may be stabilized through the coordination by MeAl(ODBP)$_2$, does not abstract the phenolic hydrogen from the monomer. In fact, the polymerization of [B] with oligo(EMA) anion formed with *t*-BuLi/MeAl(ODBP)$_2$ (1/5) in toluene at -78°C gives a block copolymer with quantitative block efficiency. Thus the monomer with a sterically protected OH group can be anionically polymerized without the need of protection as far as the reactivity of the propagating anion is modified properly. The polymers may be useful as polymeric antioxidant and as a starting material for polymeric catalysts.

References

[1] M. Szwarc, *Nature* , **178,** 1168 (1956)

[2] C. E. Schildknecht, A. O. Zoss, C. Mckinley, *Ind. Eng. Chem.*, **39**, 180 (1947).

[3] K. Ziegler, E. Holzkamp, H. Breil, H. Martin, *Angew. Chem.*, **67**, 426 (1955).

[4] G. Natta, P. Pino, P. Corradini, F. Danusso, E. Mantica, G. Mazzanti, G. Moraglio, *J. Am. Chem. Soc.*, **77**, 1708 (1955)

[5] A. H. Soum and M. Fontanille, *Makromol. Chem.*, **181,** 799 (1980)

[6] A. H. Soum, C. F. Tien, and T. E. Hogen-Esch, *Makromol. Chem., Rapid Commun.*, **4,** 243 (1983)

[7] H. Yuki, K. Hatada, *Adv. Polym. Sci.*, **31**, 1(1979)

[8] K. Hatada, T. Kitayama, K. Ute, *Prog. Polym. Sci.*, **13**, 189 (1988)

[9] O. W. Webster, *Science*, **251**, 887 (1991)

[10] H. Yasuda, H. Yamamoto, K. Yokota, S. Miyake, and A. Nakamura, *J. Am. Chem. Soc.*, **114**, 4908 (1992); H. Yasuda, H. Yamamoto, M. Yamashita, K. Yokota, A. Nakamura, S. Miyake, Y. Kai, and N. Kanehisa, *Macromolecules*, **26**, 7134 (1993)

[11] K. Hatada, K. Ute, K. Tanaka, T. Kitayama, and Y. Okamoto, *Polym. J .*, **17**, 977 (1985)

[12] K. Hatada, K. Ute, K. Tanaka, Y. Okamoto and T. Kitayama, *Polym. J .*, **18**, 1037 (1986)

[13] T. Kitayama, T. Shinozaki, E. Masuda, M. Yamamoto, and K. Hatada, *Polym. Bull.*, **20,** 505 (1988)

[14] T. Kitayama, T. Shinozaki, T. Sakamoto, M. Yamamoto, and K. Hatada, *Makromol. Chem. Suppl.*, **15**, 167 (1989)

[15] T. Kitayama, Y. Zhang, K. Hatada, *Polym. Bull.*, **32**, 439 (1994)

[16] T. Kitayama, Y. Zhang, K. Hatada, *Polym. J.*, **26**, 868 (1994)

[17] T. Kitayama, T. Hirano, K. Hatada, *Polym. J.*, **28**, 61 (1996)

[18] T. Kitayama, T. Hirano, Y. Zhang, K. Hatada, *Macromol. Symp.*, (1996), in press.

[19] H. Schlaad, A. H. E. Muller, *Macromol. Rapid Commun.*, **16**, 399 (1995)

[20] S. K. Varshney, J. P. Hautekeer, R. Fayt, R. Jerome, Ph. Teyssie, *Macromolecules,* **20**, 1442 (1990)

[21] D. G. H. Ballard, R. J. Bowles, D. M. Haddleton, S. N. Richards, R. Sellens, and D. L. Twose, *Macromolecules*, **25**, 5907 (1992)

[22] M. Kuroki, T. Watanabe, T. Aida, S. Inoue, *J. Am. Chem. Soc.*, **113**, 5903 (1991).

[23] O. W. Webster, W. R. Hertler, D. Y. Sogah, W. B. Farnham, T. V. Rajanbaub, *J. Am. Chem. Soc.*, **105**, 5703 (1983)

[24] T. Kitayama, E. Masuda, M. Yamaguchi, T. Nishiura and K. Hatada, *Polym. J.*, **24**, 817 (1992)

[25] S. Kanetaka, M. Miyamoto, T. Saegusa, *Polym. Prep. Jpn*, **39**, 220 (1990)

[26] T. Kitayama, S. He, Y. Hironaka, K. Hatada, *Polym. J.*, **27**, 314 (1995).

[27] T. Kitayama, S. He, Y. Hironaka, K. Hatada, unpublished results.

[28] T. Kitayama, S. He, T. Yabuta, K. Hatada, unpublished results.

[29] T. Kitayama, O. Nakagawa, E. Hasegawa, K. Hatada, unpublished results.

[30] T. Kitayama, S. Urbanek, T. Yanagida, K. Hatada, to be submitted to *Polym. J.*

Single Site Polymerizations of Ethylene and 1-Olefins Catalyzed by Rare Earth Metal Complexes

Eiji Ihara and Hajime Yasuda*

Department of Applied Chemistry, Faculty of Engineering,
Hiroshima University, Higashi-Hiroshima 739, Japan

Abstract: The catalytic activities of a series of metallocene lanthanide(II) complexes such as racemic $Me_2Si(2-SiMe_3-4-tBu-C_5H_2)_2Sm(THF)_2$, meso type $Me_2Si(Me_2SiOSiMe_2)(C_5H_2-tBu)Sm(THF)_2$ and C_1 symmetric $Me_2Si[2,4-(SiMe_3)_2C_5H_2][3,4-(SiMe_3)_2C_5H_2]Sm(THF)_2$ were examined for polymerization of ethylene and 1-olefins. As a consequence, catalytic activity for polymerization of ethylene increases in the order of meso > racemic > C_1 symmetric complex. Only the racemic complex exhibits good catalytic activity for polymerization of 1-olefins. The catalysis of lanthanide(III) metallocenes was also explored by using racemic $Me_2Si[2,4-(SiMe_3)_2C_5H_2]_2SmCH(SiMe_3)_2$ and C_1 symmetric $Me_2Si[2,4-(SiMe_3)_2C_5H_2][3,4-(SiMe_3)_2C_5H_2]SmCH(SiMe_3)_2$, and the C_1 symmetric complex was found to exhibit good catalytic activity for polymerization of ethylene and 1-olefins while the racemic complex is inert towards the polymerization of ethylene and 1-olefins.

INTRODUCTION

The well defined homogeneous Ziegler-Natta olefin polymerization systems include 1) highly active two component catalysts consisting of "methylalumoxane" in combination with group 4 metallocene derivative which may exhibit remarkable iso- or syndiospecificities in propylene polymerization with suitable modification of cyclopentadienyl ligand,[1-10] and 2) single component catalysts such as cationic group 4 metallocene alkyls although their activities are lower than those of the methylalumoxane containing systems.[11-20] Synthesis of polyethylene with very narrow molecular weight distribution has been attained using Group 5 $(C_5Me_5)Ta(diene)-Me$/methylalumoxane system.[21,22] More recently, neutral group 3 lanthanide metallocene hydrides or alkyls attract much attention since these systems are active even in the absence of "methylalumoxane".[23-26] For example, $[rac-Me_2Si(2-SiMe_3-4-tBu-C_5H_2)_2YH]_2$ shows the good catalysis by itself for isospecific polymerization of propylene $(M_n=4,200,\ M_w/M_n=2.32)$ and 1-olefins $(M_n>20,000,\ M_w/M_n>1.75)$.[27] $[(C_5Me_4)SiMe_2(\eta-NCMe_3)Sc(\eta-H)]_2$ also catalyzes the isospecific polymerization of 1-olefins $(M_n>6,500,\ M_w/M_n>1.5)$.[28,29] However, more simple organolanthanide complexes such as $LnH(C_5Me_5)_2$ (Ln=La, Nd) are inert for polymerization of 1-olefins although these are active for polymerization of ethylene $(M_n>590,000,\ M_w/M_n<2.03)$.[30] Among divalent

M. Kamachi · A. Nakamura (Eds)
New Macromolecular Architecture and Functions
Proceedings of the OUMS '95 Toyonaka, Osaka, Japan, 2-5 June, 1995
© Springer-Verlag Berlin Heidelberg 1996

organolanthanide species, only $Sm(C_5Me_5)_2(THF)_2$ is known to be active for polymerization of ethylene but the upper limit of the molecular weight is ca. 24,600 (M_W/M_n=2.28).

This paper deals with systematic studies on the catalytic action of racemic, meso and C_1 symmetric organolanthanide(II) and organolanthanide(III) complexes towards polymerization of ethylene and 1-olefins.

ORGANOLANTHANIDE(II) COMPLEXES WITH BULKY SUBSTITUENTS

The racemic $Me_2Si(2-SiMe_3-4-tBuC_5H_2)_2Sm(THF)_2$ (**1**) was prepared as purple crystals (mp 155.6°C) by reaction of $Me_2Si(2-SiMe_3-4-tBuC_5H_2)K_2$ with SmI_2 in THF. Single X-ray analysis clearly reveals the racemic structure (Fig. 2). Cp'(centroid)-Sm-Cp'(centroid) bite angle is 117.08(4)°, 18.6° smaller than that of $(C_5Me_5)_2Sm(THF)_2$ containing no $SiMe_2$ bridge.[31]

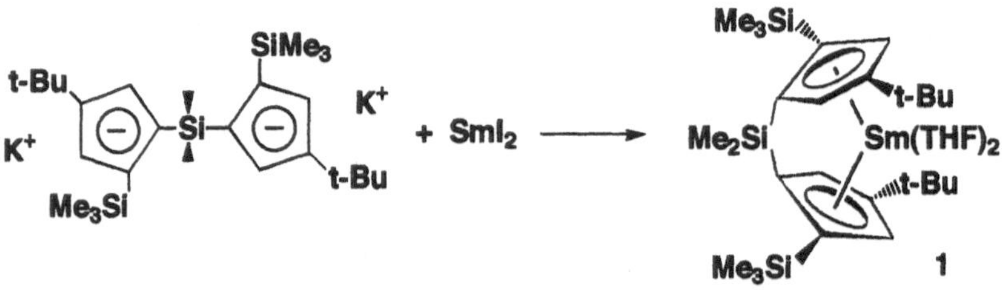

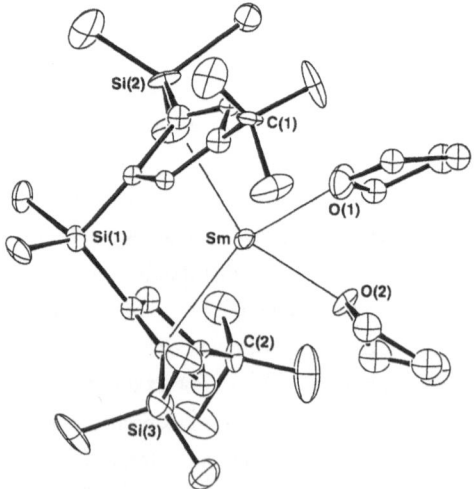

Fig. 1. Synthetic route to rac-$Me_2Si(2-SiMe_3-4-tBuC_5H_2)_2Sm(THF)_2$ (**1**)

Fig. 2. X-ray structure of rac-$Me_2Si(2-SiMe_3-4-tBuC_5H_2)_2Sm(THF)_2$ (**1**)

On the other hand, meso type complex was prepared by reaction of $Me_2Si(C_5H_3-tBu)Li_2$ with 0.5 equivalent of $ClMe_2SiOSiMe_2Cl$ followed by reaction of the resulting product with nBuLi/tBuOK to form Me$_2$Si(Me$_2$SiOSiMe$_2$)(C$_5$H$_2$-tBu)K$_2$, which was then reacted with SmI$_2$ in THF. Molecular structure of the resulting purple crystals of Me$_2$Si(Me$_2$SiOSiMe$_2$)(4-tBuC$_5$H$_2$)Sm(THF)$_2$ (**2**) (mp 168.3°C) was determined by X-ray analysis (Fig. 4). The Cp'(centroid)-Sm-Cp'(centroid) bite angle is 116.69(4)°, again ca. 19° smaller than that of the non-bridged complexes. Bis-siloxane bridged metallocene was also synthesized from the reaction between K(C$_5$H$_2$-tBu)(Me$_2$SiOSiMe$_2$)$_2$(C$_5$H$_2$-tBu)K and SmI$_2$ and the X-ray analysis reveals that bite angle of Cp'(centroid)-Sm-Cp'(centroid) is 133.4°, 16° larger as compared with that of Me$_2$Si bridged meso type organolanthanide complex. Thus introduction of two Si-O-Si bridges resulted in increased bite angle.

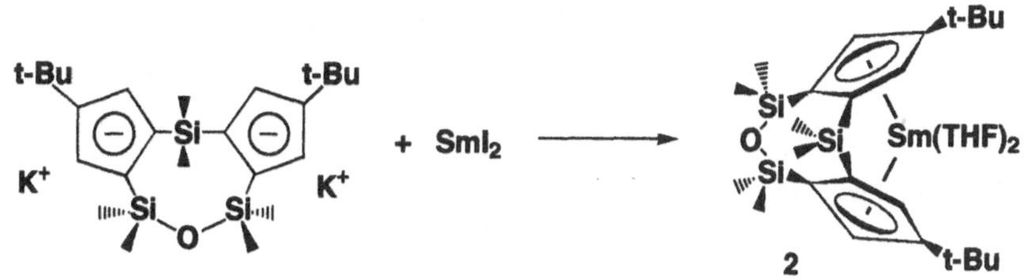

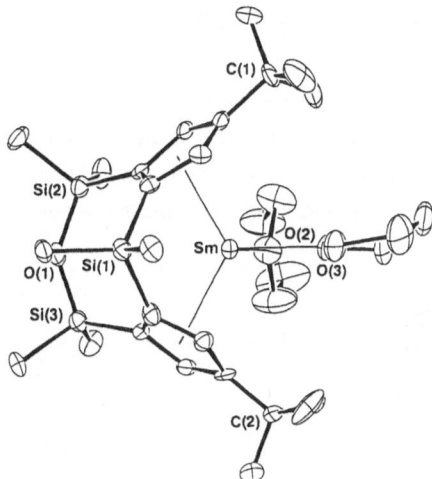

Fig. 3. Synthetic route to meso-Me$_2$Si(Me$_2$SiOSiMe$_2$)(4-tBuC$_5$H$_2$)Sm(THF)$_2$ (**2**)

Fig. 4. X-ray structure of meso-Me$_2$Si(Me$_2$SiOSiMe$_2$)(4-tBuC$_5$H$_2$)Sm(THF)$_2$ (**2**)

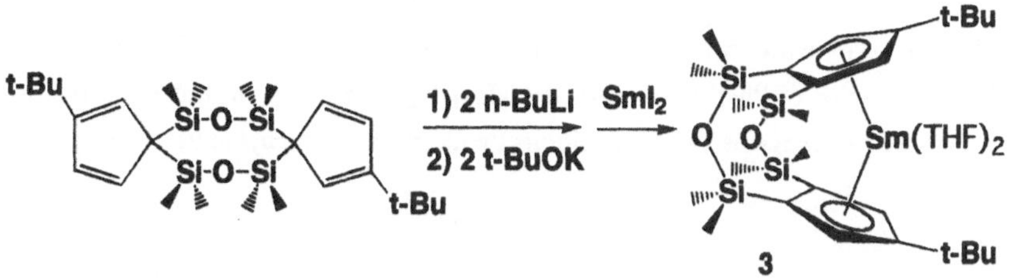

Fig. 5. Synthetic route to (Me$_2$SiOSiMe$_2$)$_2$(4-tBuC$_5$H$_2$)Sm(THF)$_2$ (**3**)

44

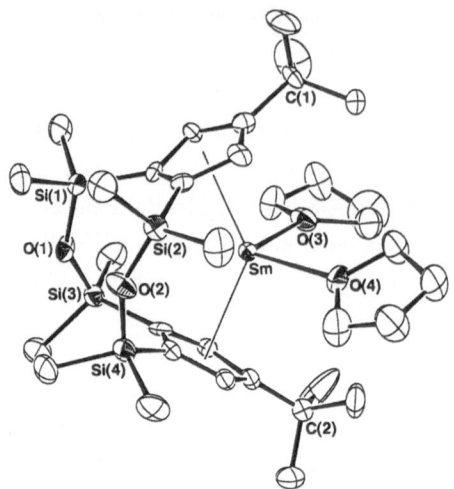

Fig. 6. X-ray structure of $(Me_2SiOSiMe_2)_2(4\text{-}tBuC_5H_2)Sm(THF)_2$ **(3)**

C_1 symmetric lanthanide metallocene, $Me_2Si[2,4\text{-}(SiMe_3)_2C_5H_2][3,4\text{-}(SiMe_3)_2C_5H_2]Sm(THF)_2$ **(4)**, was obtained as air and moisture sensitive crystals by reaction of $Me_2Si[2,4\text{-}(SiMe_3)_2C_5H_2]_2K_2$ with SmI_2 in THF. $SiMe_3$ substitution at 2 and 4-positions should migrate to 3 and 4-positions during the reaction. Thus purple crystals were obtained exclusively (mp 147.5°C). The ^1H NMR spectrum (400MHz) indicates the presence of four Cp-H groups at 50.3, 24.2, 20.8, -17.8 ppm and four $SiMe_3$ groups at 17.4, 7.8, 6.9, and -2.7 ppm. Based on the ^1H NMR spectrum, formation of single species is quite evident although X-ray analysis of the C_1 symmetric organolanthanide is now underway.

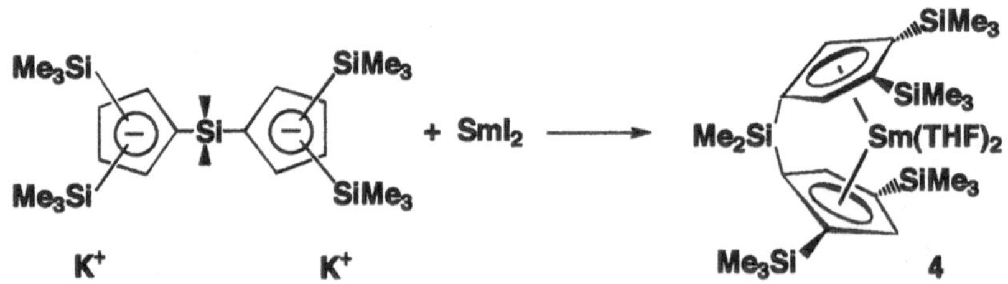

Fig.7. Synthetic route to $Me_2Si[2,4\text{-}(SiMe_3)_2C_5H_2][3,4\text{-}(SiMe_3)_2C_5H_2]Sm(THF)_2$ **(4)**

The C_{2v} symmetric complex, $Ph_2Si[3,4\text{-}(SiMe_3)_2C_5H_2]_2Sm(THF)_2$ **(5)**, was also synthesized by reaction of $Ph_2Si[3,4\text{-}(SiMe_3)_2C_5H_2]_2K_2$ with SmI_2. In this case, two $SiMe_3$ groups are forced to arrange at 3 and 4 positions by the effect of phenyl substituent.

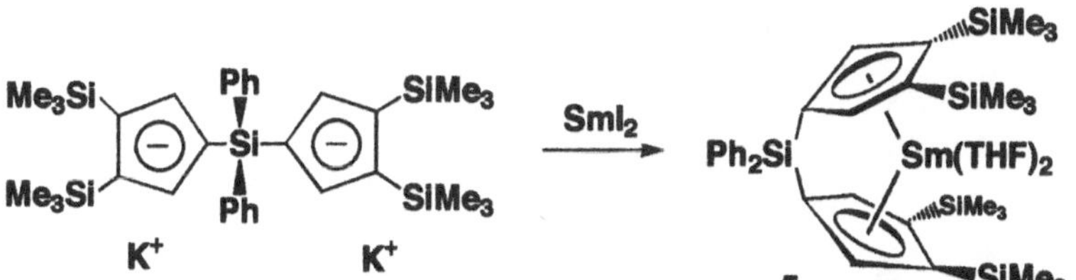

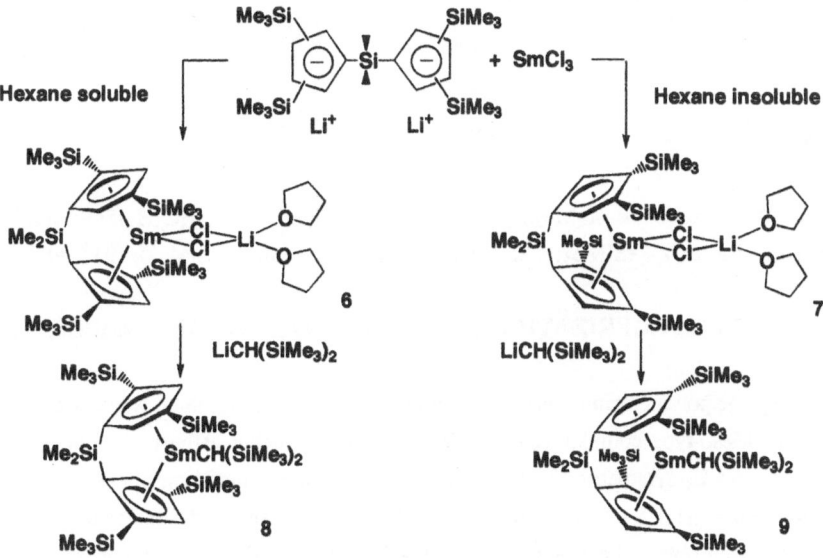

Fig. 8. Synthetic route to Ph$_2$Si[3,4-(SiMe$_3$)$_2$C$_5$H$_2$]$_2$Sm(THF)$_2$ (**5**)

ORGANOLANTHANIDE(III) COMPLEXES WITH BULKY SUBSTITUENTS

Racemic Me$_2$Si[2,4-(SiMe$_3$)$_2$C$_5$H$_2$]$_2$SmCl$_2$Li(THF)$_2$ (**6**) and C$_1$ symmetric Me$_2$Si[2,4-(SiMe$_3$)$_2$C$_5$H$_2$][3,4-(SiMe$_3$)$_2$C$_5$H$_2$]SmCl$_2$Li(THF)$_2$ (**7**) are prepared from the reaction of Me$_2$Si[2,4-(SiMe$_3$)$_2$C$_5$H$_2$]$_2$Li$_2$ with anhydrous SmCl$_3$ and the C$_1$ symmetric complex was obtained from hexane insoluble part and racemic complex from hexane soluble part. The bite angle of Cp'(centroid)-Sm-Cp'(centroid) for C$_1$ symmetric complex is 119.6° and that for racemic complex is 115.9° as revealed by X-ray analysis. Both complexes are further led to alkyl complexes by reaction with LiCH(SiMe$_3$)$_2$. After expending many efforts, we succeeded in X-ray analysis of the C$_1$ symmetric complex. Agostic interaction was observed between one of the SiMe$_3$ group attached to CH(SiMe$_3$)$_2$ and Sm metal. The bite angle of Cp'(centroid)-Sm-Cp'(centroid) is 117.9°.

Fig. 9. Synthetic route to racemic Me$_2$Si[2,4-(SiMe$_3$)$_2$C$_5$H$_2$]$_2$SmCH(SiMe$_3$)$_2$ (**8**) and C$_1$ symmetric Me$_2$Si[2,4-(SiMe$_3$)$_2$C$_5$H$_2$][3,4-(SiMe$_3$)$_2$C$_5$H$_2$]SmCH(SiMe$_3$)$_2$ (**9**)

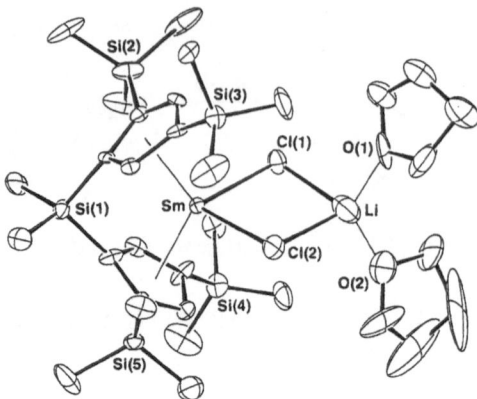

Fig. 10. X-ray structure of Me₂Si[2,4-(SiMe₃)₂C₅H₂]₂SmCl₂Li(THF)₂ (**6**)

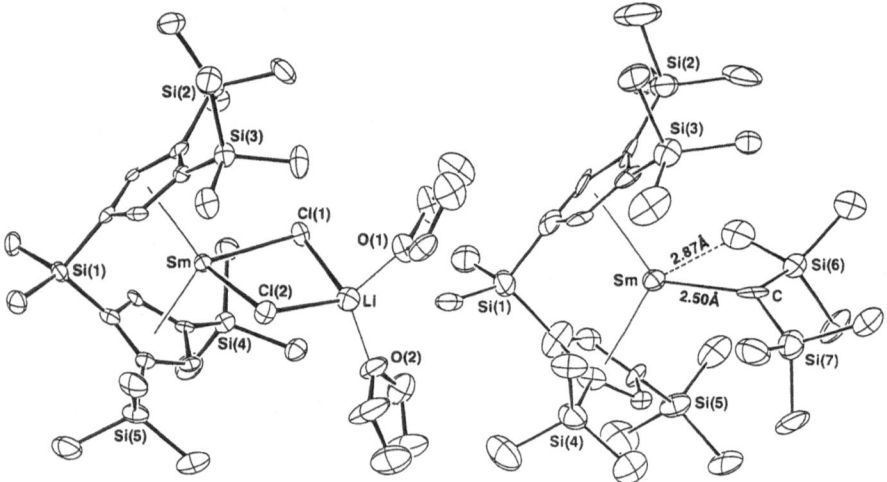

Fig. 11. X-ray structure of Me₂Si[2,4-(SiMe₃)₂C₅H₂][3,4-(SiMe₃)₂C₅H₂]SmCl₂Li(THF)₂ (**7**) (left) and Me₂Si[2,4-(SiMe₃)₂C₅H₂][3,4-(SiMe₃)₂C₅H₂]SmCH(SiMe₃)₂ (**9**) (right)

ETHYLENE POLYMERIZATION BY RARE EARTH METAL COMPLEXES

Ethylene polymerization was explored by using the resulting three types of complexes (Table 1). Meso-type complex (**3**) exhibit the highest initiating activity as compared with racemic (**1**) and C_1-symmetric (**2**) complexes as well as non-bridged metallocene derivative, (1-SiMe₃-3-tBuC₅H₃)₂Sm(THF)₂, [non-bridge-Sm(II)]. However, the molecular weight of polyethylene thus obtained are relatively small, <50,000 even when the initiator concentration was adjusted to be very low and molecular weight distributions broadened to 2.5-3.5. By contrast to these complexes, C_1-symmetric lanthanide metallocene provides the highest molecular weight of polyethylene, >1000,000 with extremely narrow molecular weight distribution although initiator efficiency is very low, 1.0-2.0%. The racemic lanthanide metallocene exhibits intermediate activity between meso-type complex and C_1 symmetric species and it gives intermediate molecular weight. Since the divalent organolanthanide should initiate the polymerization by insertion of ethylene to the two organolanthanide molecules as was found in the case of diphenylacetylene,[32] azobenzene,[33] dinitrogene,[34] stilbene[35] and butadiene[36], the initiation array may be given as shown in Scheme 1. Then the meso structure assumes the loosest packing while the C_1 symmetric structure assumes the tightest packing. Therefore meso-type complex shows the highest activity and the C_1 symmetric complex shows the lowest activity (Scheme 2).

Table 1. Ethylene polymerization initiated by divalent Sm(II) complexes

Initiators	Polymn time min	Activity (Kg-PE/mol h)	Efficiency %	$M_n/10^{-4}$	M_w/M_n
meso-tBu-Sm(II) (2)	51	46	87	1.9	3.29
	10	470	256	4.7	3.49
C_1-Sm(II) (4)	15	16	1	100.8	1.60
	30	15	2	250.2	1.80
rac-tBu-Sm(II) (1)	1	62	2	11.6	1.43
	3	139	32	35.6	1.60
C_{2v}-Sm(II) (5)	100	0	0	0	-
(SiOSi)$_2$-tBu-Sm(II) (3)	100	0.13	1	42.9	3.04
non-bridge-Sm(II)	5	6	4	1.6	2.14

Polymerization conditions, 23°C in toluene, ethylene was bubbled in 1 atm. Initiator concentration, 1.0 mM.

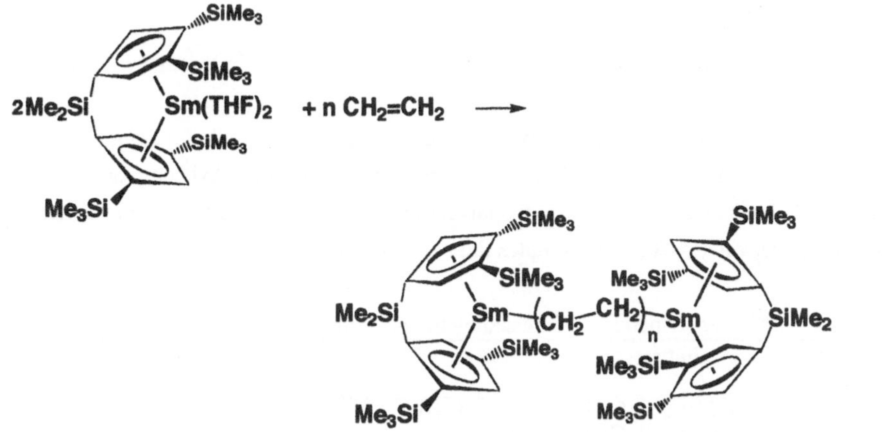

Scheme 1. Initiation mechanism for polymerization of ethylene by organolanthanide (II) complexes

48

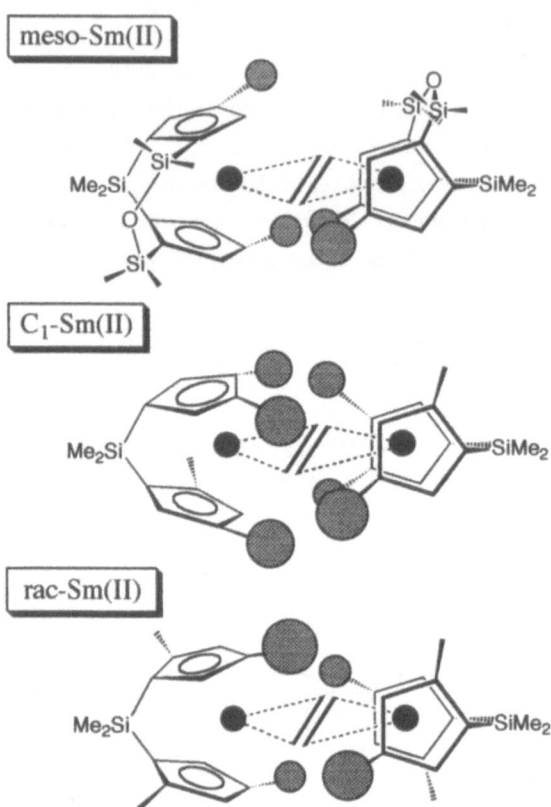

Scheme 2. Molecular packing at the initiation of polymerization of ethylene by Sm(II) species

In contrast to the high activity of divalent Sm complexes, only C1 symmetric complex shows the ethylene polymerization activity in the series of trivalent alkyl complexes. As shown in Table 2, lanthanum complex with the largest ionic radius gave the highest molecular weight polyethylene, whereas lutetium complex did not show the activity at all.

Table 2. Ethylene polymerization initiated by trivalent organolanthanide(III)

Initiators	Activity Kg/mol h	$M_n/10^{-4}$	M_w/M_n
Me$_2$Si[2(3),4-(SiMe$_3$)$_2$C$_5$H$_2$]LaCH(SiMe$_3$)$_2$	49	49.8	1.88
Me$_2$Si[2(3),4-(SiMe$_3$)$_2$C$_5$H$_2$]SmCH(SiMe$_3$)$_2$	33	41.3	2.19
Me$_2$Si[2(3),4-(SiMe$_3$)$_2$C$_5$H$_2$]YCH(SiMe$_3$)$_2$	188	33.1	1.65
Me$_2$Si[2(3),4-(SiMe$_3$)$_2$C$_5$H$_2$]LuCH(SiMe$_3$)$_2$	0	-	-

Towards the polymerization of 1-pentene and 1-hexene, only the racemic complex [rac-tBu-Sm(II) (1)] exhibits good activity, while both divalent meso and C$_1$-symmetric Sm complexes are completely inert to polymerization of those monomers. Especially noteworthy is the high isotacticity, >96%, observed in the polymerization of 1-pentene and 1-hexene. In addition to the divalent complex, C_1 symmetric alkyl complex (**9**) catalyze the polymerization of 1-olefins giving atactic poly(1-olefin)s (Table 3).

Table 3. Polymerizations of 1-olefins catalyzed by rare earth metal complexes.

Monomer	Initiator	$M_n/10^3$	M_w/M_n
1-pentene	$Me_2Si(2\text{-}SiMe_3\text{-}4\text{-}tBu\text{-}C_5H_2)_2Sm(THF)_2$ (rac) (**1**)	13	1.63
	$Me_2Si[2(3),4\text{-}(SiMe_3)_2\text{-}C_5H_2]_2YCH(SiMe_3)_2$ (C_1) (**9**)	16	1.42
	$[Me_2Si(2\text{-}SiMe_3\text{-}4\text{-}tBu\text{-}C_5H_2)_2YH]_2$ (rac)[29]	20	1.99
1-hexene	$Me_2Si(2\text{-}SiMe_3\text{-}4\text{-}tBu\text{-}C_5H_2)_2Sm(THF)_2$ (rac) (**1**)	19	1.58
	$Me_2Si[2(3),4\text{-}(SiMe_3)_2\text{-}C_5H_2]_2YCH(SiMe_3)_2$ (C_1) (**9**)	64	1.20
	$[Me_2Si(2\text{-}SiMe_3\text{-}4\text{-}tBu\text{-}C_5H_2)_2YH]_2$ (rac)[29]	24	1.75

Polymerization was carried out in toluene at 25°C.

Thus high molecular weight polyethylene was first realized by the use of rare earth metal complexes without methylalumoxane in the system. Especially noteworthy is the occurrence of living polymerizations of polar monomers such as alkyl methacrylates and alkyl acrylates by the use of organolanthanide (III)[36-38] or organolanthanide (II) complexes.[39] Thus organolanthanide complexes are active for both nonpolar and polar monomers and give the polymer of very low polydispersity.

Acknowledgment:. This work was supported by the Grant-in-Aid for Scientific Research on Priority Areas of Reactive Organometallics No. 05236104 from the Ministry of Education, Science and Culture. We thank Prof. Y. Kai and Dr. N. Kanehisa for the X-ray works.

References:

1   Kaminsky W, Külper K, Brintzinger HH, Wild FR (1985) Angew Chem Int Ed Engl 24: 507

2   Ewen JA (1984) J Am Chem Soc 106: 635

3   Ewen JA, Jones RL, Razavi A, Ferrara JD (1988) J Am Chem Soc 110: 6255

4   Erker G, Nolte R, Tsay YH, Krüger C (1989) Angew Chem Int Ed Engl 29: 629

5   Rieger B, Mu X, Mallin DT, Rausch MD, Chien JCW (1990) Macromolecules 23: 3559.

6   Chien JCW, Llinas GH, Rausch MD, Lin GY, Winter HH (1991) J Am Chem Soc 113: 8569

7   Soga K, Shiono T, Takemura S, Kaminsky W (1987) Makromol. Chem. Rapid Commun 8: 305

8   Ewen JA, Haspeslagh L, Atwood JL, Zhang H (1987) J Am Chem Soc 109: 6544

9   Ewen JA, Elder MJ, Jones RL, Haspeslagh L, Atwood JL, Bott SG, Robinson K (1991), Makromol Chem Macromol Symp 48/49: 253

10   Sinn H, Kaminsky W, Vollmer HJ, Woldt R (1980) Angew Chem Int Ed Engl 19: 390

11   Jordan RF, Bradley P, Baenziger NC, LaPointe RE (1990) J Am Chem Soc 112: 1289

12   Jordan RF (1991) Adv Organomet Chem 32: 325

13   Eshuis JJW, Tan YY, Meetsma A, Teuben JH, Renkema J, Evens GG (1992) Organometallics 11: 362.

14   Eshuis JJW, Tan YY, Teuben JH, Renkema J (1990) J Mol Cat 62: 277

15   Hlatky GG, Turner HW, Eckman RR (1989) J Am Chem Soc 111: 2728

16   Hlatky GG, Eckman RR, Turner HW (1992) Organometallics 11: 1413

17   Horton AD, Orpen AG (1991) Organometallics 10: 3910

18   Chien JCW, Tsai WM, Rausch MD (1991) J Am Chem Soc 113: 8570

19   Yang X, Stern CL, Marks TJ (1991) J Am Chem Soc 113: 3623

20   Sishta C, Hathorn R, Marks TJ (1992) J Am Chem Soc 114: 1112

21   Mashima K, Fujikawa S, Nakamura A (1993) J Am Chem Soc 115: 10990

22   Bazan GC, Donnelly SJ, Rodriguez G (1995) J Am Chem Soc 117: 2671

23   Watson PL, Parshall GW (1985) Acc Chem Res 18: 51

24   Burger BJ, Thompson ME, Cotter WD, Bercaw JE (1990) J Am Chem Soc 112: 1566

25   Jeske G, Lauke H, Mauermann H, Swepston PN, Schumann H, Marks TJ (1985) J Am Chem Soc 107: 8091

26   Yang X, Stern CL, Marks TJ (1991) Organometallics 10: 840

27   Coughlin EB, Bercaw JE (1992) J Am Chem Soc 114: 7606

28   Coughlin EB, Shapiro PJ, Bercaw JE (1992) Polym Prepr Am Chem Soc Div Polym Chem 33: 1266

29   Shapiro PJ, Cotter WD, Schaefer WP, Labinger JA, Bercaw JE (1994) J Am Chem Soc 116: 4623

30   Watson PL, Herskovitz T (1983) ACS Symp Ser 212: 459

31   Evans WJ, Grate JW, Choi HW, Bloom I, Hunter WE, Atwood JL (1985) J Am Chem Soc 107: 941

32   Evans WJ, Bloom I, Hunter WE, Atwood JL (1988) J Am Chem Soc 105: 1401

33   Evans WJ, Drummond DK, Chamberlain LR, Doedens RJ, Bott SG, Zhang H, Atwood JL (1988) J Am Chem Soc 110: 4983

34   Evans WJ, Ulibarri TA, Ziller JW (1988) J Am Chem Soc 110: 6877

35   Evans WJ, Ulibarri TA, Ziller JW (1990) J Am Chem Soc 112: 219

36   Evans WJ, Ulibarri TA, Ziller JW (1990) J Am Chem Soc 112: 2314

37   Yasuda H, Yamamoto H, Yokota K, Miyake S, Nakamura A (1992) J Am Chem Soc 114: 4908

38   Yasuda H, Yamamoto H, Yamashita M, Yokota K, Nakamura A, Miyake S, Kai Y, Kanehisa N (1993) Macromolecules 22: 7134

39   Giardello MA, Yamamoto Y, Brard L, Marks TJ (1995) J Am Chem Soc 117: 3726

40   Boffa LS, Novak BM (1994) Macromolecules 27: 6993

Non-Conjugated and Conjugated Dienes in Acyclic Diene Metathesis (ADMET) Chemistry

K. B. Wagener and T. A. Davidson

Department of Chemistry and
Center for Macromolecular Science and Engineering
University of Florida
P. O. Box 117200
Gainesville, FL 32611-7200

Abstract: Acyclic diene metathesis (ADMET) chemistry has been a subject of interest for several years now and compliments ring opening metathesis polymerization (ROMP) as a means to produce a variety of polymers via metal exchange reactions. ADMET chemistry is not a new concept, although it had been unsuccessful for a period of more than 20 years. The realization that the acidity of the catalyst influences the polymerization reaction led to the development of successful ADMET catalysts [1]. When Lewis-acid free catalysts are employed, competing reactions are eliminated, and the metathesis mechanism predominates [2]. Figure 1 shows the general reaction which occurs during an ADMET polymerization.

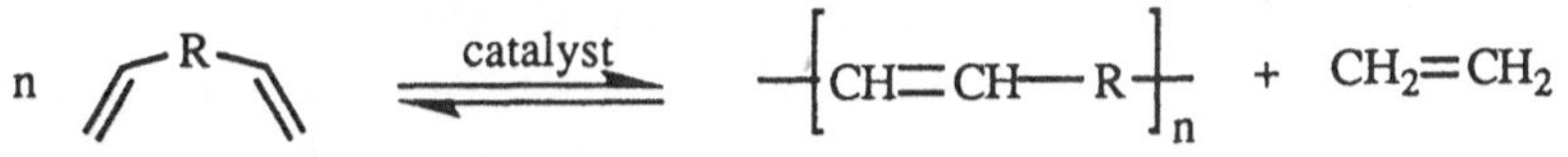

$$n \quad \underset{}{\overset{R}{/\!/\diagdown\diagup\diagdown\diagdown}} \quad \underset{}{\overset{\text{catalyst}}{\rightleftharpoons}} \quad \overbrace{}^{}\left[CH{=}CH{-}R \right]_n \quad + \quad CH_2{=}CH_2$$

Figure 1: The ADMET Polymerization.

The ADMET reaction is the latest example of step condensation polymerization where molecular weight is increased via stepwise combination of monomer units. Dienes condense via the release of a small molecule, in this case, ethylene. Due to the equilibrium nature of this reaction, continuous removal of ethylene generates high molecular weight polymer. Like other step polymerizations, ADMET forms high polymer only at near complete monomer conversion, and the molecular weight distributions of the polymer approach the statistical maximum value of 2.0.

The advent of the highly active Lewis acid-free catalyst systems developed by Schrock [3-5] in the mid to late eighties led to the discovery of the importance of using an acid-free catalyst system in ADMET reactions. Figure 2 shows the first such catalyst system employed in successful ADMET chemistry, the tungsten version of a Schrock catalyst possessing hexafluoro-t-butoxy ligands. Today, both the tungsten and molybdenum versions prove to be active catalysts for ADMET polymerization reactions [6], although the scope of available catalysts is beginning to broaden considerably. Equally of interest are the ruthenium catalyst systems produced by Grubbs [7] within the past few years (Figure 2). Using any of the three catalysts shown in Figure 2 leads to a

M. Kamachi · A. Nakamura (Eds)
New Macromolecular Architecture and Functions
Proceedings of the OUMS '95 Toyonaka, Osaka, Japan, 2-5 June, 1995
© Springer-Verlag Berlin Heidelberg 1996

polymerization mechanism that generates the necessary connections between monomer units and releases a small molecule during the polymerization cycle.

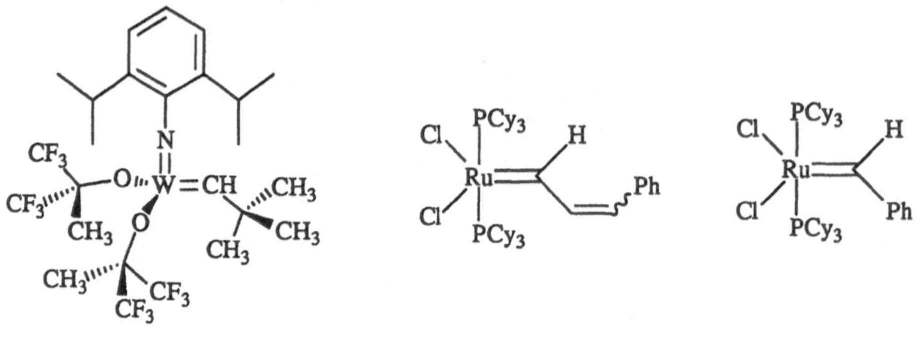

Figure 2: Catalysts used in metathesis polymerizations: A. Schrock's tungsten alkylidene. B and C. Grubbs' ruthenium catalysts.

It is interesting to note that while the overall picture for ADMET chemistry is that of a step polymerization, the cycle involving the formation of the links itself, as shown in Figure 3, appears to be more like a traditional chain polymerization. The ADMET mechanism differs considerably from ring opening metathesis polymerization (ROMP). In a ROMP reaction, only one metallacyclobutane ring is formed in each propagation step, whereas two metallocyclobutane rings are required in each ADMET propagation cycle. We continue to investigate this mechanism to further the understanding of how various factors influence ADMET chemistry.

1. NON-CONJUGATED DIENES IN ADMET CHEMISTRY

In order to study the influence of steric effects in the ADMET propagation cycle, our first investigations involved non-conjugated aliphatic dienes [8]. Figure 4 shows the initial work that was done in this regard, where a series of aliphatic dienes was examined for reactivity under typical ADMET conditions. As noted, steric effects become evident when methyl substituents preclude the formation of the first metallacyclobutane rings in the polymerization cycle. This phenomenon has been substantiated under many reaction conditions and now stands as firm evidence for steric interactions in ADMET chemistry.

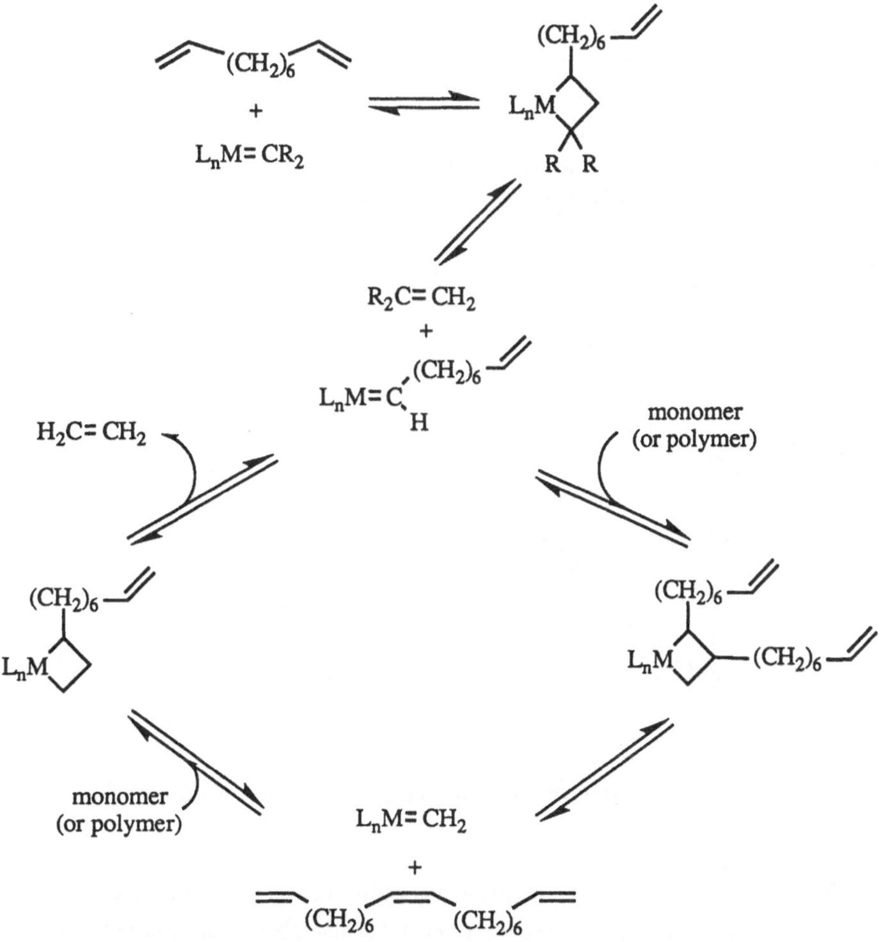

$(CH_2)_6$

$$\begin{array}{c}
\diagup\diagdown(CH_2)_6\diagdown\diagup \\
+ \\
L_nM= CR_2
\end{array} \rightleftharpoons L_nM \overset{(CH_2)_6\diagdown\diagup}{\underset{R \quad R}{\square}}$$

$$R_2C= CH_2$$
$$+$$
$$L_nM= C\overset{(CH_2)_6\diagdown\diagup}{\underset{H}{\diagdown}}$$

$H_2C= CH_2$

monomer
(or polymer)

$(CH_2)_6\diagdown\diagup$

$L_nM\,\square$

$(CH_2)_6\diagdown\diagup$

$L_nM\,\square\text{—}(CH_2)_6\diagdown\diagup$

monomer
(or polymer)

$$L_nM= CH_2$$
$$+$$

$(CH_2)_6 \qquad (CH_2)_6$

Figure 3: The ADMET polymerization cycle.

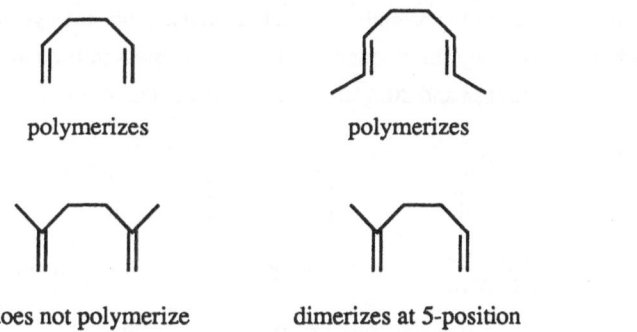

polymerizes

polymerizes

does not polymerize

dimerizes at 5-position

Figure 4: Reactivity of various aliphatic monomers towards ADMET polymerization.

This non-conjugated diene study has also led to our first analysis of the kinetics of this reaction. Investigations of the condensation rate for 1,9-decadiene have been done by measuring the rate of ethylene evolution as a function of time using the Grubbs vinyl ruthenium catalyst system. The data from our initial results is illustrated in Figure 5. There appears to be an induction period associated with ruthenium catalysis which is not evident in the case of molybdenum catalysis using Schrock's alkylidenes. This subject remains one of interest and one which is being actively pursued at the moment.

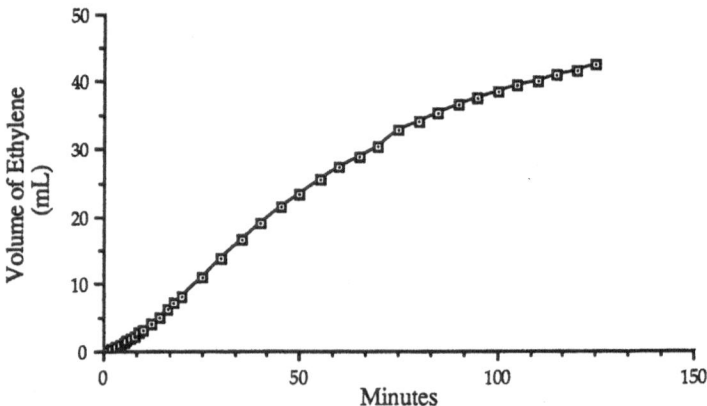

Figure 5: Evolution of ethylene with time for the polymerization of 1,9-decadiene with Grubbs' ruthenium catalyst.

The application of ADMET polymerization to non-conjugated dienes is not limited to pure aliphatic compounds. In recent years, it has become evident that a number of functional groups can be used in ADMET polymerization if proper spacing is established between the functional group and the condensing olefin [9-15]. For example, Figure 6 illustrates the reactivity for a series of ethers [9]. Divinyl ether is completely unreactive, whereas diallyl and bis-butenyl ethers show increasing rates of evolution of ethylene. This phenomenon has been observed for a number of functional groups possessing sulfur, oxygen and nitrogen and remains an area of active study.

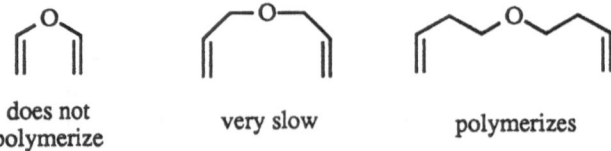

does not
polymerize

very slow

polymerizes

Figure 6: Reactivity of ethers towards ADMET polymerization.

The data collected so far for aliphatic non-conjugated dienes leads to the observations summarized in Figure 7. Aliphatic and aromatic hydrocarbons can be used in ADMET chemistry if any sterically encumbering group is kept at least beta from the metathesizing olefin position. This statement also holds true for any functional groups that might be present.

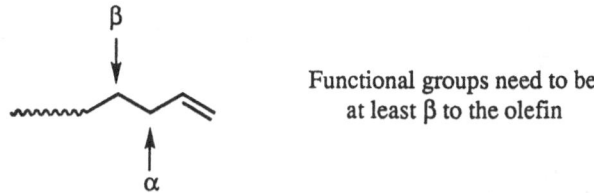

Functional groups need to be
at least β to the olefin

Functionalities which can be incorporated in ADMET polymers include:

- ketones
- esters
- carbonates
- ethers

- amines
- phosphines
- thioethers
- imides

- ionomers
- carbosilanes
- carbosiloxanes

Figure 7: The "β effect" in ADMET polymerization.

Keeping the "β effect" considerations in mind, it is possible to incorporate functionalities as shown in Figure 7, thus illustrating the breadth of the chemistry that can be accomplished using condensation metathesis reactions. We have found, however, that conjugated aliphatic dienes behave in quite a different manner. These conjugated dienes offer mechanistic challenges as yet unseen in ADMET reactions, and our results thus far are described in the following section.

2. CONJUGATED DIENES AND ACYCLIC DIENE METATHESIS POLYMERIZATION.

Our first work in the area of conjugated dienes began with para-dipropenylbenzene several years ago [16,17]. We were able to demonstrate that polyphenylene vinylene could be made via ADMET polymerization, albeit of low molecular weight (Figure 8). While this reaction was rather conventional, it demonstrated how conjugated aromatic systems could be made cleanly using room temperature chemistry. It is also noted in Figure 8 that both meta- and ortho- substituted derivatives condense to form their respective polymers, but at different reaction rates. Meta-dipropenylbenzene reacts more slowly than the para- derivative, and ortho dipropenylbenzene reacts at the slowest rate of all. While we do not fully understand the reasons for this, we suspect that it may be due to π donation back to the empty d orbital available on the metal center, thereby inhibiting its catalytic ability.

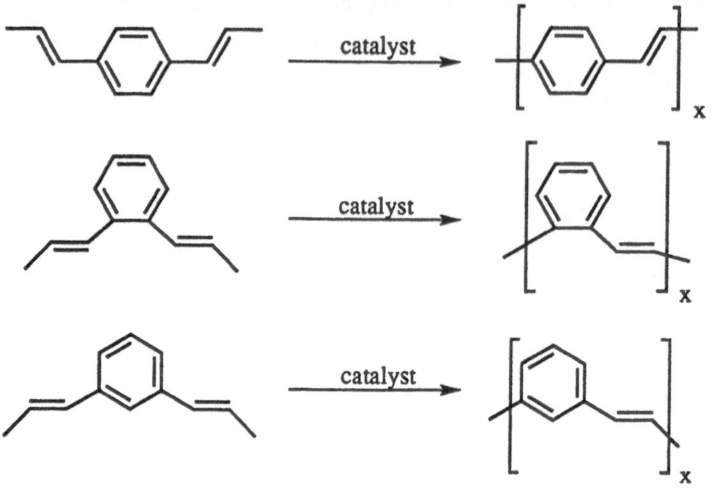

Figure 8: Poly(phenylene vinylene)s generated by ADMET polymerization.

The aliphatic conjugated dienes present even more interesting information to consider [18]. A seen in Figure 9, 2,4-hexadiene condenses quite rapidly to produce methyl terminated polyacetylene; however, its non-methyl substituted analog, 1,3-butadiene, is completely inert under these conditions. Similar observations are made when comparing 1,3,5-hexatriene and its methyl substituted analog, 2,4,6-octatriene. The methyl substituted version remains quite reactive,

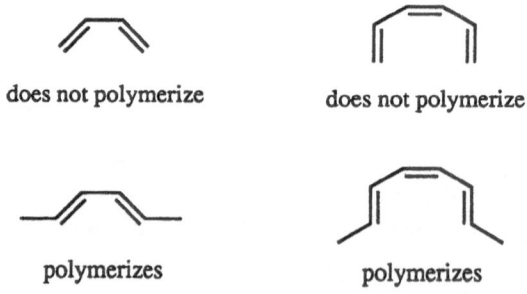

does not polymerize   does not polymerize

polymerizes   polymerizes

Figure 9: Reactivity study of conjugated monomers towards ADMET polymerization.

while the non-substituted version is completely inert. These are curious observations which require a detailed mechanistic explanation. Our first venture in this area has shown that simply adding a terminal methyl substituent, as in 1,3-pentadiene, also leads to an inert polymerization system. Likely explanations for this lack of reactivity might be associated with tight π-bond interactions with the metal center, thereby completely obviating catalytic activity. Our investigations are continuing in this area to study this chemistry in greater depth.

ACKNOWLEDGMENTS

We would like to acknowledge the National Science Foundation, Shell Development Company, and Air Products and Chemicals, Inc. for their generous support of this research, and Dr. John Anderson for his contributions in preparation of this manuscript.

REFERENCES

1   Ivin KJ (1983) Olefin Metathesis, Academic Press, New York
2   Lindmark-Hamberg M, Wagener KB (1987) Macromolecules 20:2949
3   Schrock RR (1986) J Organomet Chem 300:249
4   Feldman J, Davis WM, Schrock RR (1989) Organometallics 8:2266
5   Schrock RR, DePue RT, Feldman J, Schaverian JC, Dewan JC, Liu RH (1988) J Am Chem Soc 110:1423
6   Konzelman J, Wagener KB (1995) Macromolecules 28:4686
7   Nguyen ST, Grubbs RH (1993) J Am Chem Soc 115:9858
8   Konzelman J (1993) PhD Dissertation, University of Florida
9   Brzezinska K, Wagener KB (1992) Macromolecules 25:2049
10 O'Gara JE, Portmess JD, Wagener KB (1993)Macromolecules 26:2837
11 Smith DW, Wagener KB (1991) Macromolecules 24:6073
12 Smith DW, Wagener KB (1993) Macromolecules 26:3533
13 Patton JT, Wagener KB, Forbes MDE, Myers TL, Maynard HD (1992) Polym Prepr (Am Chem Soc, Div Polym Chem) 33:1070
14 Wagener KB, Patton JT, Boncella JM (1992) Macromolecules 25:3862
15 Wagener KB, Patton JT (1993) Macromolecules 26:249
16 Wolf A, Wagener KB (1991) Polym Prepr (Am Chem Soc, Div Polym Chem) 32:535
17 Tao D (1994) PhD Dissertation, University of Florida
18 Tao D, Wagener KB (1994) Macromolecules 27:1281

Design and Synthesis of New Substituted Polyacetylenes

Toshio Masuda and Yoshihiko Misumi

Department of Polymer Chemistry, Kyoto University, Kyoto 606-01, Japan.

Abstract: The present paper deals with recent studies on the metathesis polymerization of substituted acetylenes: (a) the living polymerization by $MoOCl_4$-based catalysts; and (b) the synthesis and properties of poly(diphenylacetylenes). Several acetylenes ($ClC\equiv C$-n-C_6H_{13}, $HC\equiv C$-t-Bu, etc.) undergo living polymerization with $MoOCl_4$–n-Bu_4Sn–EtOH (1:1:1). It is noteworthy that *tert*-butylacetylene provides a stereoregular living polymer. Whereas phenylacetylene does not polymerize in a living fashion, its derivatives with bulky ortho groups (e.g., CF_3) undergo virtually ideal living polymerization. The $MoOCl_4$–Et_3Al–EtOH catalyst also effects living polymerization. Diphenylacetylenes with various groups (e.g., p-Me_3Si, p-t-Bu, p-PhO, and p-carbazyl) polymerize in high yields with $TaCl_5$–n-Bu_4Sn. The formed polymers are yellow solids totally soluble in toluene and thermally very stable (M_w 1×10^6–3×10^6). Some of them are more permeable to gases than is poly(dimethylsiloxane).

Introduction

Substituted acetylenes can be polymerized by suitable transition-metal catalysts (eq. 1; see ref. 1, 2 for reviews and ref. 3–7 for recent typical papers). The produced polymers have alternating double bonds along the main chain and various groups as the side chains. The polymerization of substituted acetylenes by group 5 and 6 transition metal catalysts is thought to proceed via metal carbenes; i.e., a metal carbene reacts with an acetylene to give a metallacyclobutene, ring opening of which regenerates a metal carbene (eq. 2). This mechanism resembles that of ring-opening metathesis polymerization (ROMP) of cycloolefins.[8-10]

Polymerization of RC≡CR'

$$ RC\equiv CR' \xrightarrow{\text{catalyst}} \begin{array}{c} \\ \left(C=C\right)_n \\ | \quad | \\ R \quad R' \end{array} \quad \left(\begin{array}{c} \text{Substituted} \\ \text{Polyacetylene} \end{array} \right) \quad (1) $$

Propagation mechanism

$$ \text{wwC}=M \quad C\equiv C \quad \longrightarrow \quad \text{wwC}-M \quad \longrightarrow \quad \text{wwC} \quad M \quad (2) $$
$$ (M: \text{metal}) \qquad\qquad\qquad | \quad | \qquad\qquad || \quad || $$
$$ C=C \qquad\qquad C-C $$

M. Kamachi · A. Nakamura (Eds)
New Macromolecular Architecture and Functions
Proceedings of the OUMS '95 Toyonaka, Osaka, Japan, 2-5 June, 1995
© Springer-Verlag Berlin Heidelberg 1996

As a typical function of substituted polyacetylenes, one can point out gas-separation membrane based on their high gas permeability.[11,12] Further, various functions of substituted polyacetylenes are being developed extensively. Their examples include separation of ethanol/water mixture by pervaporation,[13] and third-order nonlinear optical properties.[14]

Herein we introduce our recent studies on the polymerization of substituted' acetylenes: (a) living polymerization by $MoOCl_4$-based catalysts; and (b) synthesis and properties of poly(diphenylacetylenes).

Living Polymerization by $MoOCl_4$-Based Catalysts

Living polymerization is one of the most useful means to control both molecular weight and molecular weight distribution (MWD) of polymers. Recently, many living processes have been developed in not only anionic but also various types of polymerizations.[15] Whereas the study on the living ROMP of cycloolefins has recently made great progress,[8,9] there have been only a few reports on the living polymerization of substituted acetylenes; e.g., [o-(trimethylsilyl)phenyl]acetylene/Schrock carbene[3] and several substituted acetylenes/$MoOCl_4$–n-Bu_4Sn–EtOH.[16-20]

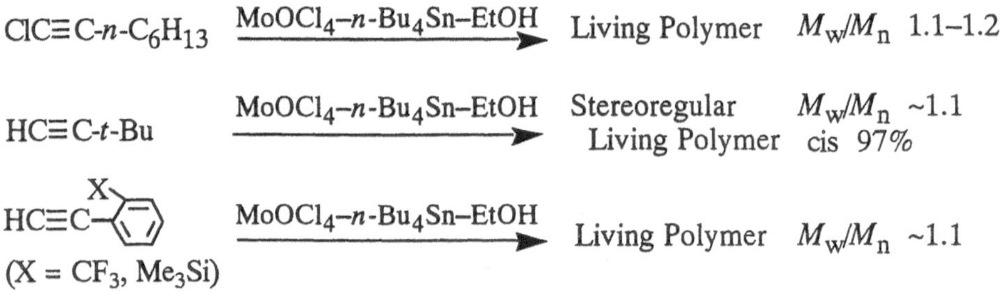

$ClC{\equiv}C\text{-}n\text{-}C_6H_{13}$ $\xrightarrow{MoOCl_4\text{–}n\text{-}Bu_4Sn\text{–}EtOH}$ Living Polymer M_w/M_n 1.1–1.2

$HC{\equiv}C\text{-}t\text{-}Bu$ $\xrightarrow{MoOCl_4\text{–}n\text{-}Bu_4Sn\text{–}EtOH}$ Stereoregular M_w/M_n ~1.1
Living Polymer cis 97%

$HC{\equiv}C\text{-}$〈X〉 $\xrightarrow{MoOCl_4\text{–}n\text{-}Bu_4Sn\text{–}EtOH}$ Living Polymer M_w/M_n ~1.1
(X = CF_3, Me_3Si)

Scheme 1. Living polymerization of substituted acetylenes
by $MoOCl_4$–n-Bu_4Sn–EtOH (1:1:1)

Scheme 1 shows acetylenic monomers that undergo living polymerization by $MoOCl_4$–n-Bu_4Sn–EtOH catalyst. 1-Chloro-1-octyne provides a living polymer whose polydispersity ratio is 1.1–1.2.[16] Quite interestingly, tert-butylacetylene produces a stereoregular living polymer, that is, a polymer having 97% cis and polydispersity ratio of 1.1.[17] Such stereospecific living polymerizations are rare, even when one considers all the known polymerization mechanisms. Further, phenylacetylenes having ortho substituents also polymerize in a living fashion; examples of such monomers include phenylacetylenes with o-CF_3, o-Me_3Si, and o-Me_3Ge.[18,19] This result is noteworthy because phenylacetylene itself does not give a living polymer with this catalyst.

Figure 1 shows MWD curves of the polymers formed from 1-chloro-1-octyne.[16] The polymerizations have been carried out with $MoOCl_4$-based catalysts in toluene at 30 °C, and the monomer feed has been supplied three times every 5 min. In the polymerization by $MoOCl_4$–n-Bu_4Sn, the polymer molecular weight increases progressively, but the MWDs are not narrow. In contrast, $MoOCl_4$–n-Bu_4Sn–EtOH (mole ratio 1:1:1) decreases the polydispersity ratio to 1.1–1.2. Here the polymer molecular weight increases progressively with each further supply of monomer. These results manifest that this polymerization is a living polymerization.

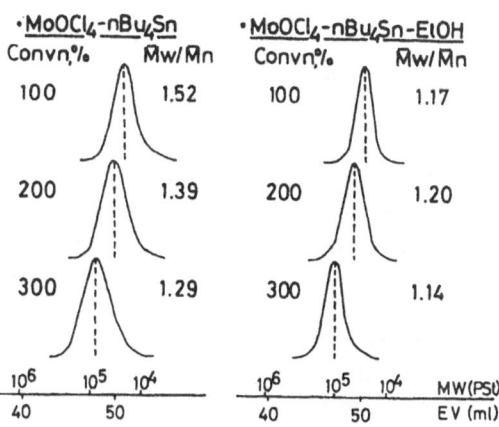

Figure 1. MWD curves of poly(1-chloro-1-octyne)s formed with $MoOCl_4$-based catalysts (in toluene, 30 °C, $[M]_0 = 0.10$ M, $[MoOCl_4] = 10$ mM).

tert-Butylacetylene has turned out to undergo not only living but stereospecific polymerization in the presence of the $MoOCl_4$-based ternary catalyst.[17] The spectrum in Figure 2a is the ^{13}C NMR of the poly(*tert*-butylacetylene) obtained with $MoOCl_4$ alone. The signal due to the methyl groups splits into two peaks a_1 and a_2, which are attributable to the cis and trans structures, respectively, of the main chain. The spectrum in Figure 2b is the one for the polymer formed with the $MoOCl_4$-based living catalyst. Here only the

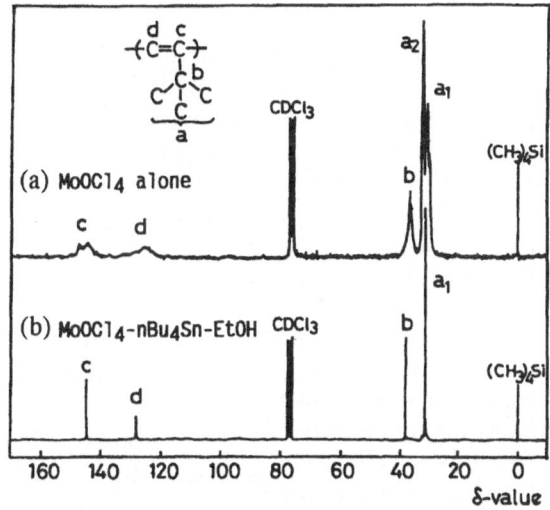

Figure 2. ^{13}C NMR spectra of poly(*tert*-butylacetylene) obtained with $MoOCl_4$-based catalysts (polymerized in toluene at 0 °C).

sharp a_1 peak is seen, and hence it is concluded that this polymer has not only a narrow MWD but all-cis structure.

It has been examined how the steric control of poly(*tert*-butylacetylene) occurs in the polymerization by $MoOCl_4$–n-Bu_4Sn.[20] The polymer obtained just after all the monomer has been consumed in this polymerization has all-cis

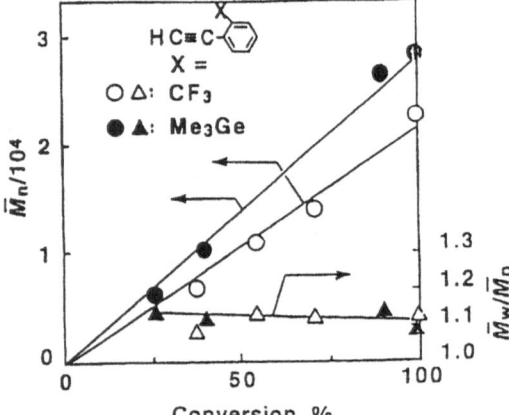

$$H-C\equiv C-t\text{-}Bu \xrightarrow{\text{Mo cat}} \begin{array}{c} \text{cis} \\ \\ \text{trans} \end{array} \quad \left(C=C \right)_n \xrightarrow[\text{EtAlCl}_2]{\text{Acid} \left(\text{e.g., MoOCl}_4, \right)} \left(C=C \right)_n$$

(3)

structure according to ^{13}C NMR (eq. 3). However, isomerization to trans proceeds on further standing the polymerization system. Eventually it became clear that various acids including $MoOCl_4$ cause geometric isomerization. These results lead to a conclusion that only cis structure is formed in the propagation reaction, but that isomerization to trans takes place afterwards depending on the kind of catalysts.

Interestingly, phenylacetylenes having bulky ortho substituents also undergo living polymerization.[18,19] For example, the polymerizations of o-CF_3- and o-Me_3Ge-phenylacetylenes proceed smoothly without an induction phase to reach 100% conversion finally. In these polymerizations, the M_n of polymer increases in direct proportion to monomer conversion, while the polydispersity ratio remains as small as 1.1 (Figure 3). Thus these monomers undergo virtually ideal living polymerization.

Table I summarizes how the ortho substituent of phenylacetylene affects the living polymerization by the $MoOCl_4$-based ternary catalyst.[18] When the ortho substituent is none or small, the polymerizations are non-living. On the other hand, the phenylacetylenes that have medium-sized ortho substituents polymerize in a living fashion, though the MWDs are

Figure 3. Polymerization of phenylacetylenes with bulky ortho substituents by $MoOCl_4$–n-Bu_4Sn–EtOH (1:1:1) (in toluene, 30 °C, $[M]_0 = 0.10$ M, $[MoOCl_4] = 10$ mM).

Table I. Polymerization of ortho-Substituted Phenylacetylenes by $MoOCl_4$–n-Bu_4Sn–EtOH (1:1:1) [a]

Ortho-Substituent		M_w/M_n	Livingness
Bulkiness	Example		
Small	H, F	> 2.0	No
Medium	CH_3, Cl, iPr	1.2–1.3	Yes
Large	CF_3, Me_3Ge	~ 1.1	Yes

[a] In toluene, 30 °C, $[M]_0 = 0.10$ M, $[MoOCl_4] = 10$ mM.

not very narrow. In contrast, the monomers having very bulky groups exhibit excellent living nature with small polydispersity ratios of about 1.1. Thus the steric effect of ortho substituents is very important to achieve living polymerization.

Recently Et3Al has been examined as a cocatalyst other than n-Bu4Sn in the polymerization of o-CF3-phenylacetylene by MoOCl4-based ternary catalysts[21] (Figure 4). The n-Bu4Sn-containing catalyst system gives a polymer whose polydispersity ratio is 1.06. On the other hand, the Et3Al-involving counterpart produces a polymer, which has a polydispersity ratio as small as 1.02. In this case, anisole is preferable to toluene as solvent, which is probably due to the strong reducing ability of Et3Al.

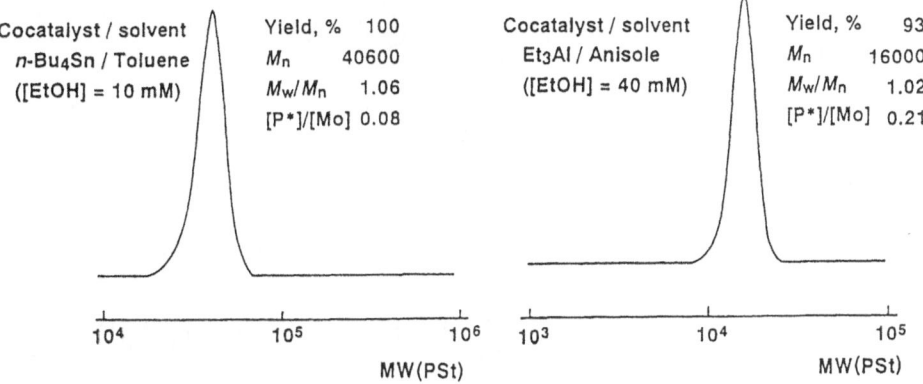

Figure 4. Polymerization of o-CF3-phenylacetylene by MoOCl4-based catalysts (30 °C, 1 h; $[M]_0 = 0.20$ M, $[MoOCl_4] = [Cocat] = 10$ mM).

Table II shows various MoOCl4-based catalysts effective for living polymerization.[21] Not only n-Bu4Sn but also Et3Al, Et2Zn, and n-BuLi are useful as cocatalysts in the MoOCl4-based system. The ratios of catalyst components and polymerization solvents should be properly chosen. Quite interestingly, the polydispersity ratios of poly(o-CF3-phenylacetylene) are as small as 1.06 to 1.02. The values for poly(1-chloro-1-octyne) were 1.1–1.2. Thus various cocatalysts are available for achieving the living polymerization of substituted acetylenes.

Table II. MoOCl4-Based Catalysts for the Living Polymerization of o-CF3-phenylacetylene

Catalyst	Solvent	M_w/M_n
MoOCl4–n-Bu4Sn–EtOH (1:1:1)	toluene	1.06
MoOCl4–Et3Al–EtOH (1:1:4)	anisole	1.02
MoOCl4–Et2Zn–EtOH (1:1:3)	anisole	1.02
MoOCl4–n-BuLi–EtOH (1:1:2)	anisole	1.02

Synthesis and Properties of Poly(diphenylacetylenes)

In the presence of $TaCl_5$-based catalysts, diphenylacetylene (DPA) forms a polymer, which is thermally very stable but insoluble in any solvent[22] (Scheme 2). There has been a tendency regarding polymer solubility that polyacetylenes having two identical alkyl groups in the repeat unit are insoluble, whereas polyacetylenes having methyl and a long alkyl are soluble in organic solvents. This is attributable to the difference in polymer surface area between these two types of polymers. Hence a working hypothesis is possible that if a bulky substituent is introduced into one of the phenyl groups of DPA, then the polymer may become soluble.

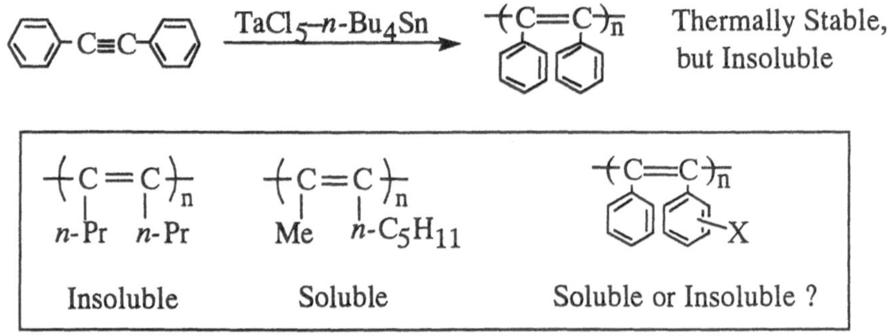

Scheme 2. Research background for the study on substituted poly(DPAs)

The objectives of this study include synthesis of new, soluble polymers from DPA derivatives, characterization of the polymers, and development of polymer functions. The monomers used were DPAs that have Me_3Si, t-butyl, n-butyl, phenoxy, and carbazyl groups at para or meta position.

Scheme 3 shows results for the polymerization of various DPAs by $TaCl_5$–n-Bu_4Sn. p-Me_3Si-DPA polymerizes in high yield up to 85%.[23] As was hoped, the polymer is totally soluble in many common solvents such as toluene and $CHCl_3$. Quite interestingly, the molecular weight of the polymer reaches about two million. The m-Me_3Si derivative also provides a soluble polymer, whose molecular weight exceeds one million. Although Nb and Ta belong to the same group in the periodic table, the Nb counterpart does not polymerize these monomers.

Among alkyl-containing DPAs, the t-Bu derivative forms a polymer in about 85% yield.[24] The polymer is totally soluble in toluene and $CHCl_3$ and its weight-average molecular weight determined by gel permeation chromatography (GPC) reaches 3.6 million. An interesting point is that the absolute M_w value determined by light scattering is smaller and about a half of the value by GPC. The n-Bu-containing polymer is also totally soluble, and its M_w value is about one million.

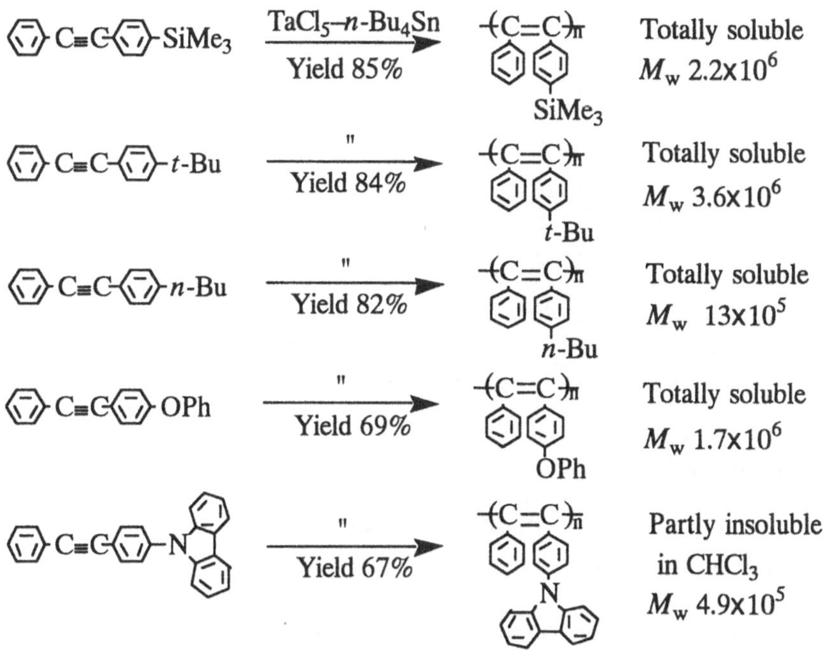

Scheme 3. Polymerization of ⟨◯⟩-C≡C-⟨◯⟩-X by TaCl$_5$–n-Bu$_4$Sn(1:2) (in toluene, 80 °C, 24 h; [M]$_o$ 0.50 M, [TaCl$_5$] = 20 mM; M_w by GPC)

The phenoxy derivative also achieves a high polymer yield of around 70% despite the presence of an ether linkage.[25]   The polymer molecular weight is higher than one million, and as high as for other derivatives.   The carbazyl-containing monomer gives a mostly soluble polymer, whose molecular weight is about 5x10[5].[26]

Figure 5 illustrates UV–visible spectra of poly(DPAs) measured in THF.   Every polymer shows two absorption maxima at about 370   and 430 nm, whose molar absorptivities are 4000 to 6000 M^{-1} cm^{-1}.   The cut-off wavelengths of the absorptions are about 500 nm irrespective of the para substituents.   These spectra correspond to the yellow color of the polymers.

Figure 6 depicts TGA curves for various poly-

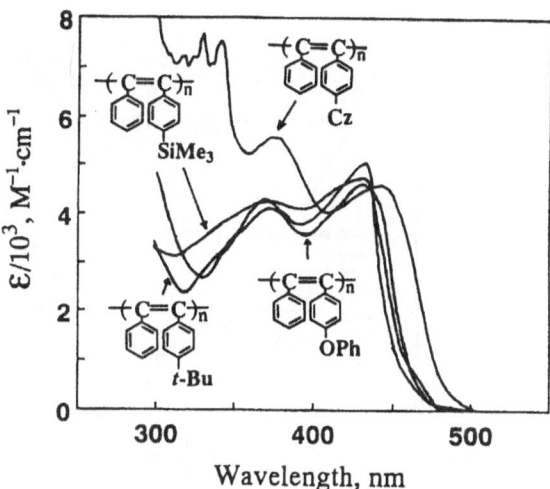

Figure 5. UV-visible spectra of poly(diphenylacetylenes) (in THF).

(DPAs) measured in air. Poly(phenylacetylene) begins to lose weight at a temperature as low as 200 °C. In contrast, poly(DPA) keeps weight up to 500 °C, and is more stable than any other substituted polyacetylenes. The onset temperatures of weight loss for the poly(DPA)s having substituents are usually between 400 and 500 °C, indicating high thermal stability.

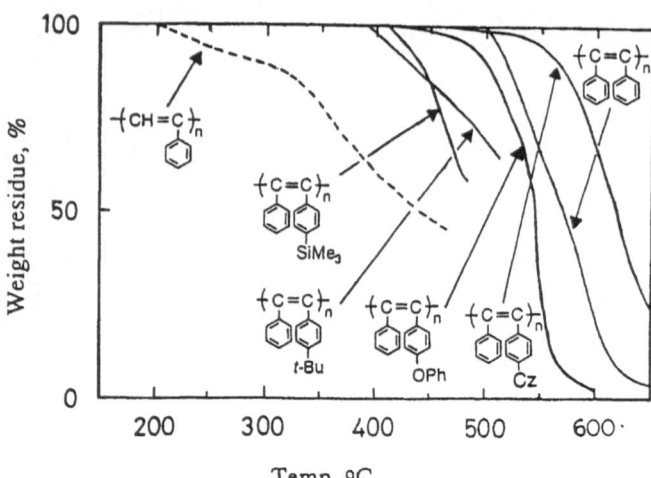

Figure 6. TGA curves of poly(diphenylacetylenes) (in air, heating rate 10 °C/min).

Gas permeability of the present polymers were examined as a polymer function (Figure 7). Poly[1-(trimethylsilyl)-1-propyne] shows a P_{O_2} value about ten times higher than that of poly(dimethylsiloxane) and is more permeable to oxygen than are any other polymers. In the present study we have found that the poly(DPAs) having round-shaped ring substituents are about twice as permeable to oxygen as is poly(dimethylsiloxane). It is noteworthy that the shape of ring substituents in poly(DPA) plays an important role in gas permeability.

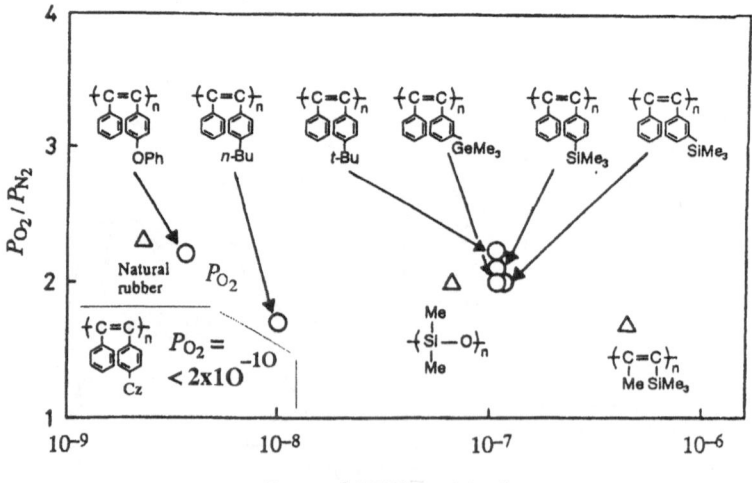

Figure 7. Plot of oxygen permeability coefficient (P_{O_2}) vs. separation factor (P_{O_2}/P_{N_2}) for poly(diphenylacetylenes) and a few other polymers (25 °C).

Table III. Polymerization of Diphenylacetylenes and Polymer Properties

Polymn by		TaCl$_5$–n-Bu$_4$Sn	(yield >50%)		
M_w / 10^3	–	2200	3600	1700	490
In toluene	insoluble	soluble	soluble	soluble	mostly sol.
T_0, °C^a	500	420	380	420	470
P_{O_2}, barrer^b	–	1100	1100	37	< 2

a Onset temp. of weight loss in TGA (in air).

b P_{O_2}: Oxygen permeability coefficient; 1 barrer = 1×10^{-10} cm^3(STP)·cm /(cm^2·s·cmHg)

Table III summarizes the results of poly(DPA)s. Not only poly(DPA) but also its derivatives can be obtained by the polymerization using TaCl$_5$–n-Bu$_4$Sn. The polymer molecular weights reach a few million. Most of the substituted poly(DPAs) are soluble in toluene and CHCl$_3$. The onset temperatures of weight loss for poly(DPA) and its derivatives are usually 400–500 °C, showing high thermal stability. The oxygen permeability coefficients, P_{O2}, of these polymers having round-shaped ring substituents are about 1000 barrers, nearly twice that of poly(dimethylsiloxane).

Thus poly(DPA) derivatives are a new class of polyacetylenes which exhibit interesting properties, and hence further development of their unique functions is expected.

Acknowledgment

The authors thank coworkers whose names are shown in the cited references. Thanks are also due to Ms. Satoko Nakatsuka for technical assistance. This work was partly supported by Grant-in-Aid for Scientific Research on Priority Areas (07216236) from the Ministry of Education, Science, and Culture, Japan.

References

1) H. Shirakawa, T. Masuda, and K. Takeda, *The Chemistry of Triple-Bonded Functional Groups*, Supplement C2, S. Patai, Ed., Wiley, Chichester, 1994, Chap. 17.

2) T. Masuda and H. Tachimori, *J. Macromol. Sci.-Pure Appl. Chem.*, **A31**, 1675 (1994).

3) R. R. Schrock, S. Luo, N. C. Zanetti, and H. H. Fox, *Organometallics*, **13**, 3396 (1994).

4) H.-J. Lee, S.-J. Kang, H.-K. Kim, H.-N. Cho, J.-T. Park, and S.-K. Choi, *Macromolecules*, **28**, 4638 (1995).

5) A. Furlani, C. Napoletano, M. V. Russo, A. Camus, and N. Marsich, *J. Polym. Sci., Part A, Polym. Chem.*, **27**, 75 (1989).

6) M. Tabata, Y. Inaba, K. Yokota, and Y. Nozaki, *J. Macromol. Sci., Pure Appl. Chem.*, **A31**, 465 (1994).

7) Y. Kishimoto, P. Eckerle, T. Miyatake, T. Ikariya, and R. Noyori, *J. Am. Chem. Soc.* **116**, 12131 (1994).

8) R. R. Schrock, *Acc. Chem. Res.*, **23**, 158 (1990).

9) B. M. Novak, W. Risse, and R. H. Grubbs, *Adv. Polym. Sci.*, **102**, 47 (1992).

10) D. S. Breslow, *Prog. Polym. Sci.*, **18**, 1141 (1993).

11) A. C. Savoca, A. D. Surnamer, and C.-F. Tien, *Macromolecules*, **26**, 6211 (1993).

12) H. Odani and T. Masuda, *Polymers for Gas Separation*, N. Toshima, Ed., VCH, New York, 1992, Chap. 4.

13) T. M. Aminabhavi, R. S. Khinnavar, S. B. Harogoppad, U. S. Aithal, Q. T. Nguyen, and K. C. Hansen, *J. Macromol. Sci., Rev.*, **C34**, 139 (1994).

14) T. Wada, T. Masuda, and H. Sasabe, *Mol. Cryst. Liq. Cryst.*, **247**, 139 (1994).

15) T. Aida, *Prog. Polym. Sci.*, **19**, 469 (1994).

16) T. Masuda, T. Yoshimura, and T. Higashimura, *Macromolecules*, **22**, 3804 (1989).

17) M. Nakano, T. Masuda, and T. Higashimura, *Macromolecules*, **27**, 1344 (1994).

18) T. Mizumoto, T. Masuda, and T. Higashimura, *Macromol. Chem. Phys.*, **196**, 1769 (1995).

19) T. Masuda, K. Mishima, J. Fujimori, M. Nishida, H. Muramatsu, T. Higashimura, *Macromolecules*, **25**, 1401 (1992).

20) T. Masuda, H. Izumikawa, Y. Misumi, and T. Higashimura, *Macromolecules*, submitted.

21) H. Kaneshiro, S. Hayano, and T. Masuda, to be published.

22) A. Niki, T. Masuda, and T. Higashimura, *J. Polym. Sci., Part A, Polym. Chem.*, **25**, 1553 (1987).

23) K. Tsuchihara, T. Masuda, and T. Higashimura, *Macromolecules*, **25**, 5816 (1992).

24) H. Kouzai, T. Masuda, and T. Higashimura, *J. Polym. Sci., Part A, Polym. Chem.*, **32**, 2523 (1994).

25) H. Tachimori, T. Masuda, H. Kouzai, and T. Higashimura, *Polym. Bull.*, **32**, 133 (1994).

26) H. Tachimori and T. Masuda, *J. Polym. Sci., Part A, Polym. Chem.*, in press.

Precursor Method for Polymeric LB Films

Masa-aki Kakimoto*, Aiping Wu and Yoshio Imai

Department of Organic and Polymeric Materials,
Tokyo Institute of Technology, Meguro-ku, Tokyo 152, Japan

Abstract: First, the general concept of the "Precursor Method" for the preparation of poly-meric Langmuir-Blodgett (LB) films that possess no long alkyl chain between film layers, and the fabrication of electroluminescence (EL) devices using poly(p-phenylenevinylene) (PPV) LB film are described.   The preparation of PPV LB film was carried out by the same procedure as that used to make polyimide LB films via precursor LB films of polysulfonium salts.   EL devices which consisted of single-layer (ITO/PPV/Mg-Ag), double-layer (ITO/HTL/PPV/Mg-Ag), and three-layer (ITO/HTL/PPV/ETL/Mg-Ag) structures were prepared, where triphenylamine containing polyimide LB film and vapor-deposited Alq3 were used as the HTL and ETL. respectively.   EL efficiency increased with increasing the number of layers.

1. Introduction for the precursor method

Technologies for manufacturing molecular assemblies have advanced rapidly in the past decade.   The Langmuir-Blodgett (LB) technique is one of the most useful candidate methods for the assembly of organic molecules,   because the fabrication of mono-molecular thin films using the LB technique is possible.   The molecules used for LB films must be amphiphilic and are typically low molecular weight long chained aliphatic acids (fatty acids).   This class of LB films are regarded as the standard and, consequently, they have a long history.   On the other hand, when these films are used in the fabrication of novel devices, many shortcomings arise from their inherent thermal and mechanical instabilities.   Polymeric LB films have been stud-ied for two main reasons.   One of them was to determine if mono-molecular layers of polymers could be fabricated, and if so, what was the nature of such polymeric LB films, and second, to improve the physical shortcomings of low molecular weight LB films as mentioned above.

We have investigated polyimide LB films.[1)]   Polyimides are well known for their high degree of thermal stability and superior electrical properties.   Because of their infusibility and insolubility in organic solvents, it is difficult to process polyimides, so processing them into films is done through their soluble precursor polymers (polyamic acids).   A preparative method of polyimide LB films is shown in Eq. 1.   Polyamic acids 1 are mixed with long alkyl amines 4 that introduce hydrophobic alkyl chains required to form the amphiphilic   structure 2. The LB films of polyamic acid long alkyl amine salts 2 are prepared by the usual LB techniques, which consist of the preparation of a monolayer at the air-water interface and the deposition of this layer onto a substrate.   The obtained LB films of 2 are dipped into a mixed solution of

M. Kamachi · A. Nakamura (Eds)
New Macromolecular Architecture and Functions
Proceedings of the OUMS '95 Toyonaka, Osaka, Japan, 2-5 June, 1995
© Springer-Verlag Berlin Heidelberg 1996

acetic anhydride and pyridine to convert the LB films into polyimide **3**. Alternatively, the conversion can be achieved by heating the precursor LB films at 300 °C for a short period of time. The most interesting characteristic of polyimide LB films is that they do not have a long alkyl chain in their structure, which is unlike most LB films. Thus, the obtained polyimide LB films have excellent thermal stability and solvent resistance as high as commercially available polyimide thick films. We have applied polyimide LB films to devices such as solar battery cells,[2] photomemory devices working in the photon mode,[3] and liquid crystalline optical cells where they were used as an aligning layer for a liquid crystalline cell.[4]

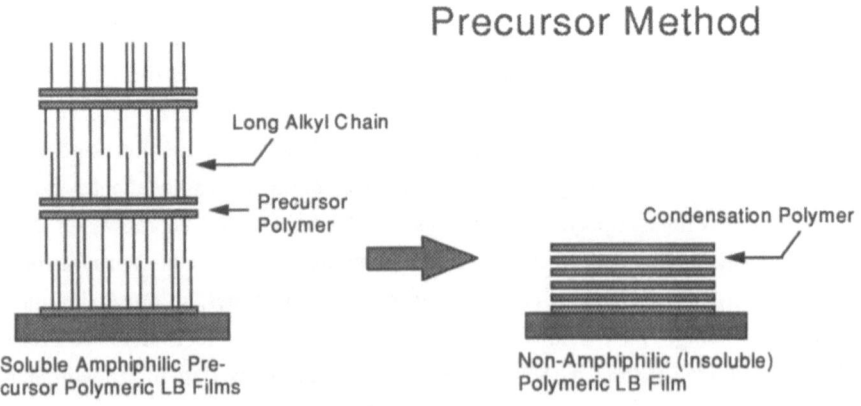

Polyamic acid (PAA) **1**

PAA Alkyl amine Salt **2**

(1)

Polyimide (PI) **3**

Precursor Method

Long Alkyl Chain

Precursor Polymer

Condensation Polymer

Soluble Amphiphilic Pre-cursor Polymeric LB Films

Non-Amphiphilic (Insoluble) Polymeric LB Film

Fig. I Concept of Precursor Method

General aspects of the "Precursor Method" for polymeric LB films are illustrated in Fig. I.

The goal is the preparation of polymeric LB films having no alkyl chain between the layers. This class of polymeric LB films is expected to have higher thermal and mechanical stability

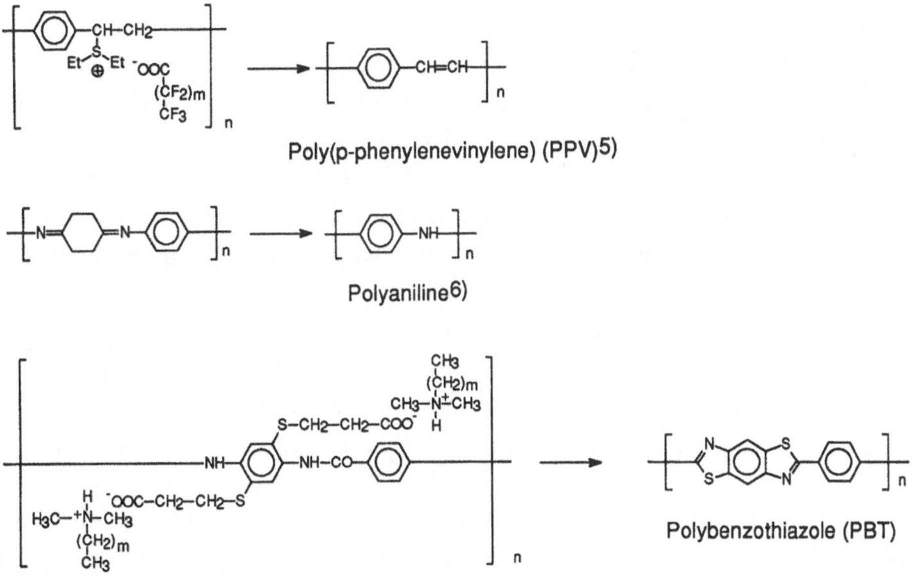

Poly(p-phenylenevinylene) (PPV)[5)]

Polyaniline[6)]

Polybenzothiazole (PBT)

Scheme 1

compared with usual amphiphilic polymeric LB films. The preparation of such LB films directly from the polymer solutions is fairly difficult, because the structure of the polymers is not amphiphilic and they are sometimes insoluble in organic solvents. Thus, we designed the precursors that are amphiphilic and soluble in organic solvents. Polyamic acid alkyl amine salts are one class of precursor polymers, which can be used for polyimide LB films. Scheme 1 shows other precursor polymers applicable to the "Precursor Method" and the final LB polymers. Poly(p-phenylene vinylene)s (PPVs) are used not only as conducting polymers but also as emitting layer in electroluminescence devices.[7)]

In this paper, recent results of the preparation of electroluminescence (EL) devices using poly(p-phenylenevinylene) (PPV) LB film.

2. Preparation of electroluminescence (EL) devices

Recently, research concerning electroluminescence (EL) of fluorescent π-conjugated polymers has become a rapidly developing field, and there is a strong possibility that polymeric EL devices will find commercial use in various display applications, especially, in large-area displays.[7-15)]

Electroluminescence is the direct conversion of electrical energy into light without the involvement of any intermediate processes. In π-conjugated polymers, EL results from the recombination of an injected electron in the conduction band with an injected hole in the valence band under applied voltage, which generates either singlet or triplet excitons. The radiative decay of singlet excitons emits light with a wavelength corresponding to the energy difference between the two bands.

Electroluminescent activity of fluorescent π-conjugated polymers, such as, poly(p-phenylene vinylene) (PPV), poly(p-phenylene) (PPP),[16] polyaniline (PA)[17] and poly(p-alkylthiophene)[18] (PAT) have been demonstrated, since J. H. Burroughes et al firstly fabricated polymeric EL devices based on PPV thin film in 1990.[7] Among these π-conjugated polymers, PPV are mostly used in polymeric EL devices, wherein, full color display in PPV based EL devices can be controlled by chemical tuning of the band gap and EL efficiency has been improved by using substituted PPVs or segmented PPVs.

However, the low EL efficiency of polymeric EL devices is still a subject of intense study. Recently, hetero-structural polymeric EL devices, containing additional charge transporting layers between the light emitting layer and the electrodes, has been fabricated and improvement of EL efficiency has been demonstrated. In these references, polysilane[19] or N,N'-diphenyl-N,N'-bis(3-methylphenyl)-1,1'-biphenyl-4,4'-diamine (TPD) vapor-deposited film have been used as the hole transporting material and 2-(4-biphenyl)-5-(4-tert-butylphenyl)-1,3,4-oxadiazole (PBD) dispersed in PMMA or Alq3 dopant have been used as the electron transporting material.

We have concentrated our work on the development of hetero-structural polymeric EL devices based on PPV Langmuir-Blodgett (LB) films. By this LB technique,[20, 21] highly ordered thin films with precise control of film thickness could be realized allowing for the fabrication of ultra thin hetero-structural devices of controlled thickness.

Experimental
Materials

The light emitting PPV LB film was prepared by the thermal conversion of the PPV precursor sulfonium salt LB multilayer, which possesses amphiphilicity by mixing poly(p-xylene diethylsulfonium chloride) with sodium perfluorononanoate. After removal of perfluorononanoic alkyl long-chain by annealing at 200 ^{0}C for 3 hrs, uniform PPV LB multilayer could be obtained with monolayer thickness of 0.34 nm. A novel hole transporting material, triphenylamine containing polyimide, was firstly utilized in our work. The hole transporting polyamide LB multilayer with monolayer thickness of 0.5 nm was prepared by the chemical imidization of precursor polyamic acid-alkyl amine salt LB films. A fluorescent metal chalet complex, 8-hydroxyquinoline aluminum (Alq3), was used as the electron transporting material, which was deposited by vacuum evaporation at pressures of about 10^{-5} Torr.

EL devices

As shown in Fig. 1, our unique EL device consists of a hole injection electrode, which is semi-transparent indium-tin oxide (ITO) film partially coated on the glass substrate; a hole transporting polyimide and a light emitting PPV LB film layers, where the LB films are deposited successively on the front surface of the ITO anode; an electron transporting vapor-deposited Alq3 film, and an electron injection electrode, which is a low work function Mg-Ag alloy cathode rectifying contact on the organic layer. The Mg-Ag alloy electrode were later deposited by vacuum evaporation at pressures of about 10^{-6} Torr. Three kinds of EL devices with single-layer, double-layer and three-layer structures have been fabricated.

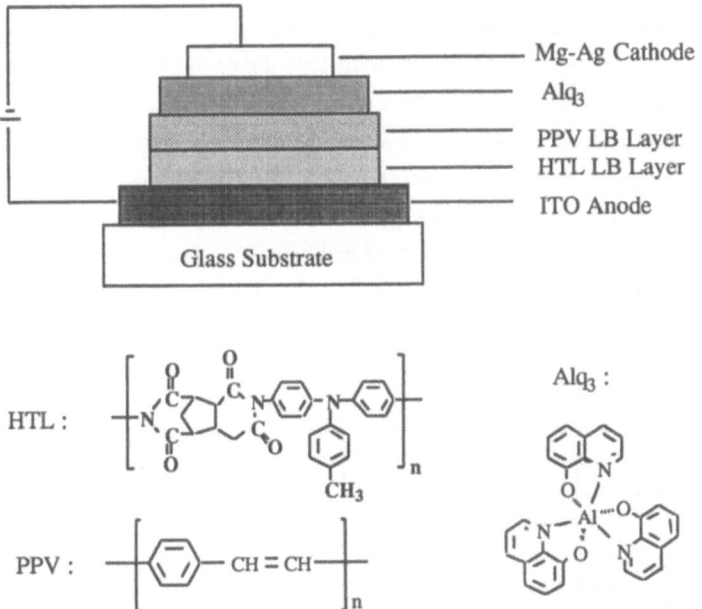

Fig. 1 Structure of EL devices and chemical structure of each layer

Measurements

The deposition of LB films were performed using a San-Esu FSD-20 LB deposition apparatus. The UV-visible spectrum was measured with a Hitachi U-4300 spectrometer. The excitation, photoluminescent (PL) and EL spectra were recorded using a Jasco FP-77 spectrofluorometer. The value of ionization potential Ip (HOMO) was determined from the measurement of photoelectron emission by AC-1, and the value of electron affinity Ea (LUMO) was assumed to be the difference between Ip and the optical absorption edge. EL devices were driven at room temperature under nitrogen atmosphere. The light output (luminance) was measured with a Topcon BM-8 Luminance meter.

Results and discussion

The absorption characteristics of PPV LB film in the UV-visible region show a maximum absorption near 450 nm, corresponding to the π-π* transition moment of the π-conjugated PPV. A longer extended π-conjugation length when compared to that of a conventional cast film was implied by the red-shifted (30 nm) absorption in the PPV LB film.[21] Additionally, due to photoexcitation, strong PL comparable to the conventionally cast PPV films was observed for the PPV LB film. Meanwhile, the hole transporting polyimide does not emit and has no influence to the optical property of PPV. However, the electron transporting Alq3 emits light with the peak of about 520 nm, which overlaps the emission of PPV film.

The energy levels of three materials is shown in Fig. 2. Comparing the energy level of the hole transporting layer (HTL)-polyimide and the emitting layer-PPV, the highest occupied molecular orbital (HOMO) level of HTL is higher than that of the PPV by 0.25 eV. Hence, a lowered energy barrier between the anode may improve the injection of the hole from the ITO anode. Meanwhile, the lowest unoccupied molecular orbital (LUMO) level of the PPV is lower than that of the HTL by 1.11 eV. Due to the high energy barrier between the HTL and the PPV layers, the injected electron from the cathode may be confined in the PPV layer. Comparing the energy level of PPV and the electron transporting layer (ETL)-Alq3, the HOMO level of Alq3 is a little bit lower than that of PPV, which indicated that the injected hole may be confined in the PPV layer.

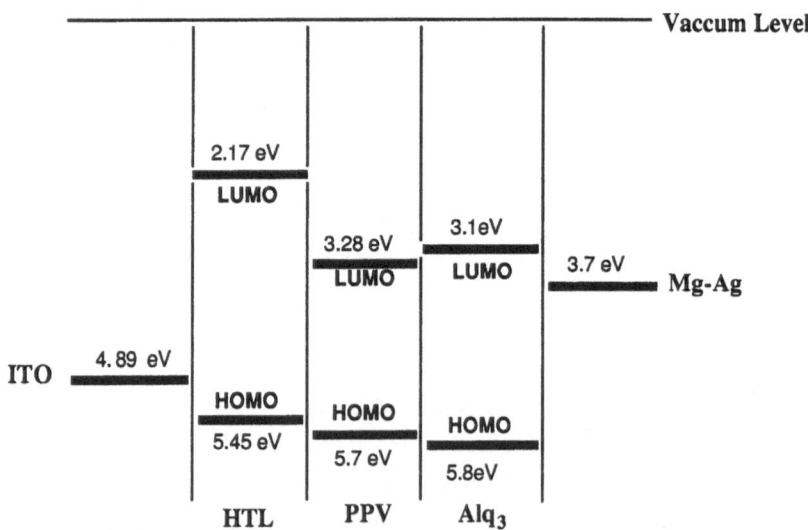

Fig. 2   Energy level of each layer

The single layer device was fabricated with only PPV LB film, which is denoted as ITO/PPV/Mg-Ag. When a forward bias was applied to this device, yellow-green emissions could be observed at a darkened room. Its EL spectrum, recorded at room temperature in air, was similar to the PL spectrum of PPV. This similarity demonstrated that the same excited state was produced either by recombination of the injected electron and hole in the double-layered device or by photo-excitation of the PPV film.

The double-layer device was fabricated with a hole transporting polyimide LB layer and PPV LB layer, which is denoted as ITO/HTL/PPV/Mg-Ag. On a forward bias, bright yellow-green emissions could be observed even under normal lighting conditions. The EL spectra was also similar to the PL spectra of PPV. This indicated that the emission comes from PPV layer. Furthermore, the EL and PL spectra did not change in the presence of the HTL, which demonstrated that there was no exciplex formed between the polyimide HTL and the PPV.

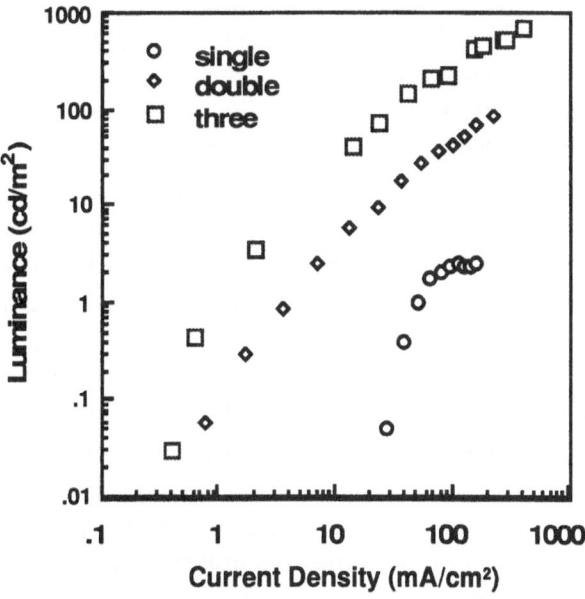

Fig. 3 Relationship between light out put and current density

In addition, EL performance was found to be highly dependent on the HTL at constant PPV thickness. The introduction of the HTL caused a notable reduction in the drive field, the relation between the threshold voltage and the polymer thickness, typically 10^6 V/cm, indicating an improvement in charge injection by use of an HTL. Furthermore, when the applied voltage was kept below 20 V, the light output (luminance) increased approximately linearly with increasing current density, as shown in Fig. 3. This fact implied that the light emission originated from the recombination of the injected charges (holes and electrons) coming from the electrodes. In the double-layer devices, the maximum luminance reached 100 cd/m^2, which is over 10 times higher than that of the single-layer device. Similarly, EL efficiencies, are improved significantly in the double-layer devices, which showed the highest EL efficiency of 0.02 lm/W for the case of an ITO/HTL(81L)/PPV(50L)/Mg-Ag device, as shown in Table 1. The dependence of the EL efficiency on the thickness of the polyimide HTL suggests that charge injection and transportation efficiency can be improved by increasing the thickness of the HTL. Comparing the energy-levels shown in Fig. 2, the function of the HTL in the LED devices was confirmed as a marked improvement of hole injection efficiency from the ITO anode and the confinement of electrons at the interface between the HTL and PPV layers.

Table 1. Data of EL devices

Structure of Devices	Voltage (V)	J (mA/m^2)	L (cd/m^2)	η_{max}(lm/W,%)*
PPV (50 L)	8	65.2	1.69	0.10
HTL(80L)/PPV(50L)	9	13.6	5.9	1.52
HTL(31L)/PPV(20L)/Alq3(30nm)	7	14.0	41.1	13.2

*Efficiency is calculated by following equation:

η(lm/W) = πL[cd/m^2] /Vj[W/m^2]

η: EL efficiency,   L: luminance,   J: current density

The three-layer device was fabricated with a hole transporting polyimide LB layer, PPV LB layer and an electron transporting Alq3 layer, which is denoted as ITO/HTL/PPV/ETL/Mg-Ag. EL performance of the three-layer device was almost the same as that of the two-layer device, but higher emission brightness of 722 cd/m^2 and higher EL efficiency of 0.13 lm/W were achieved.

In the presence of the electron transporting Alq3, the EL efficiency was improved significantly, which can be attributed to both the improvement of electron injection and the confinement of the hole. Although the EL spectrum of the three-layer device is similar to that of PPV, it is still difficult to say that light emission come from only PPV, as the PL spectrum of Alq3 overlaps that of PPV. But we may conclude that light emission occurred at the interface between the PPV and Alq3 layers.

Furthermore, the EL spectrum in the three-layer device depended on the thickness of either PPV and Alq3 layer. The More detail investigations of this phenomenon are under way.

Conclusion

Three kinds of PPV LB film based EL devices, single-layer, double-layer and three-layer structures, were fabricated, wherein a triphenylamine containing polyimide LB layer and a vapor-deposited Alq3 layer were used as the HTL and ETL, respectively. Significant improvement of EL efficiency was realized in hetero-structure devices. Especially, in three-layer device, higher EL efficiency of 0.13 lm/W was obtained, which indicated that either the injection or the confinement of charges was improved by incorporation of our charge transporting layers.

References

1 a) Suzuki M, Kakimoto M, Konishi T, Imai Y, Iwamoto M, Hino T (1986) Chem Lett 395. b) Kakimoto M, Suzuki M, Konishi T, Imai Y, Iwamoto M, Hino T (1986) Chem Lett 823

2 a) Nishikata Y, Morikawa A, Kakimoto M, Imai Y, Hirata Y, Nishiyama K, Fujihira M (1989) J Chem Sci Chem Commun 1772. b) Nishikata Y, Morikawa A, Kakimoto M, Imai Y, Nishiyama K, Fujihira M (1990) Polymer J 22:593

3 Yokoyama S, Kakimoto M, Imai Y (1994) Thin Solid Fims 242:183

4 Nishikata Y, Morikawa A, Takiguch Y, Kanemoto A, Kakimoto K, Imai Y (1988) Jpn J Appl Phys 27: L1163

5 Nishikata Y, Kakimoto M, Imai Y (1988) J Chem Soc Chem Commun 1040

6 Nishikata Y, Suwa T, Kakimoto M, Imai Y (1992) Thin Solid Films 210/211:390

7 Burroughes JH, Bradley DDC, Brown AR, Marks RN, Mackay K, Friend RG, Burns PL, Holmes AB (1990) Nature 347:539

8 Bradley DDC (1993) Synthetic Metals 54:401

9 Braun D, Heeger AJ, Kroemer H (1991) J Electronic Materials 20(11):9

10 Greenham NC, Moratti SC, Bradley DDC, Friend RH, Holmes AB (1993) Nature 365: 628

11 Yang Z, Sokolik I, Karasz FE (1993) Macromolecules 26:1188

12 Burn PL, Holmes AB, Kraft A, Bradley DDC, Brown AR, Friend RH (1992) J Chem Soc Chem Commun :32

13 Burn PL, Holmes AB, Kraft A, Bradley DDC, Brown AR, Friend RH, Gymer RW (1992) Nature 356:47

14 Friend R, Bradley D, Holmes A (1992) Physics World :42

15 Brown AR, Bradley DDC, Burroughes JB, Friend RH, Greenham NC, Burn PL, Holmes AB, Kraft A (1992) Appl Phys Lett. 61(23): 2793

16 Grem G, Leditzky G, Ullrich B, Leising G (1992) Synthetic Metals 51:383

17 Bsiesy A, Nicolau YF, Ermolieff A, Muller F, Gaspard F (1995) Thin Solid Films 255:43

18 Ohmori Y, Uchida M, Muro K, Yoshino K (1991) Solid State Communications 80(8):605

19 Kido J, Nagai K, Okamoto Y, Skotheim T (1991) Appl. Phys Lett 59(21):2760

20 Kakimoto M, Suzuki M, Konishi T, Imai Y, Iwamoto M, Hino T (1986) Chem Lett :823

21 Wu A, Yokoyama S, Watanabe S, Kakimoto M, Imai Y, Araki T, Iriyama K (1994) Thin Solid Film 244:750

Polyethyleneimine Derivatives as Nucleic Acid Model and Interaction with DNA

Yoshiaki INAKI, Takehiko WADA

Department of Applied Chemistry,

Faculty of Engineering,

Osaka University, Suita, Osaka 565, Japan

Abstract: Polyethyleneimine derivatives having nucleic acid bases and hydrophilic amino acids such as homoserine and serine were prepared as nucleic acid models. The polymers were found to interact with DNA accompanied by induction of conformational change. The induced conformation of DNA by interaction with the polymer containing uracil and homoserine (PEI-Hse-Ura) was concluded as super triple helical structure. The formation of the polymer complex, DNA : PEI-Hse-Ura, was found to be affected by the presence of metal ions such as Ca^{2+} and Cu^{2+}.

INTRODUCTION

DNA is known to form a double helical structure by specific hydrogen bondings between complementary nucleic acid bases as shown in Figure 1. The effects of chemical structure of DNA on the specific interaction have been studied using synthetic nucleic acid analogs.[1-6] Polyethyleneimine derivatives containing adenine and thymine were prepared by the grafting on the polymer by the activated ester method (Figure 2).[7] These polymers interact each other to form the polymer complexes by base pairing between adenine and thymine.[8-12] The polymers also interacted with polynucleotides such as poly (A) or poly (U).[8] The polyethyleneimine derivatives of nucleic acid base, however, were hardly soluble in water at neutral pH. Therefore, the interaction studies were restricted to the water-ethylene glycol mixed solution system.

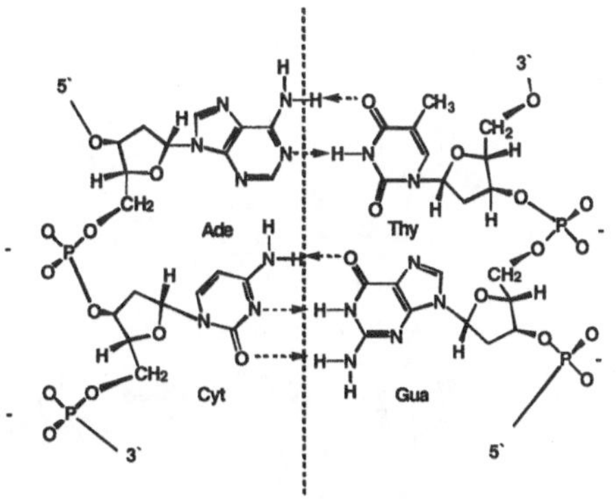

Figure 1. Base Pairs in DNA.

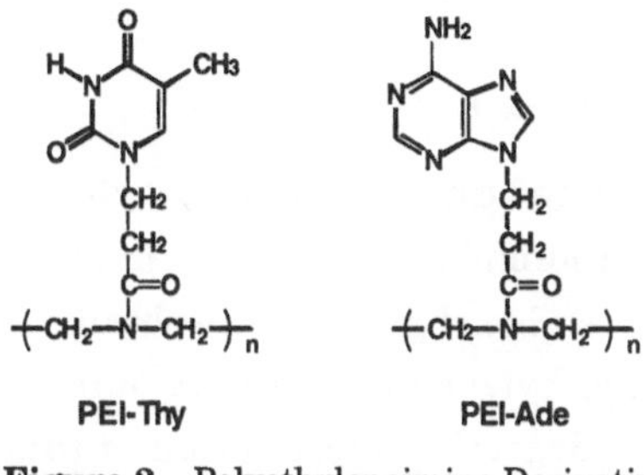

Figure 2. Polyethyleneimine Derivatives.

M. Kamachi · A. Nakamura (Eds)

New Macromolecular Architecture and Functions

Proceedings of the OUMS '95 Toyonaka, Osaka, Japan, 2-5 June, 1995

© Springer-Verlag Berlin Heidelberg 1996

Water soluble nucleic acid models were prepared using hydrophilic amino acids as a spacer such as serine and homoserine. These polymers made it possible for us to study the interactions with water soluble polynucleotides and nucleic acids in aqueous solution.[13-16]   This paper deals with the polyethyleneimine derivatives having uracil as a nucleic acid base and homoserine as the spacer (PEI-Hse-Ura) (Figure 3).   This polymer was found to form polymer complexes with natural nucleic acids, DNA, RNA, and polynucleotides.   Conformational changes of DNA by the interaction with the polyethyleneimine were studied by CD spectra.   Effects of additives such as metal ions, Cu^{2+} and Ca^{2+} on the formation of the polymer complex were also studied.

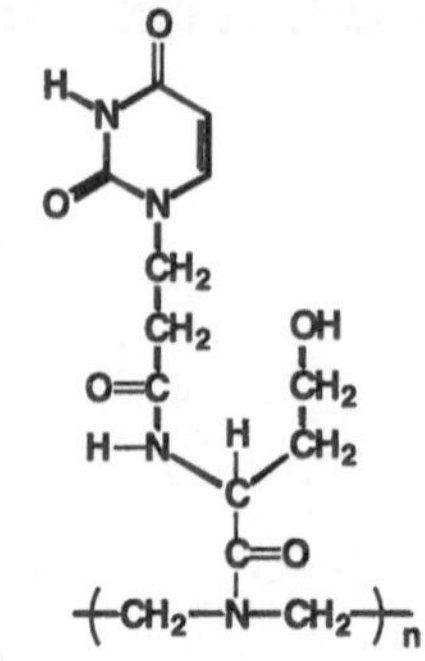

PEI-Hse-Ura

Figure 3.

EXPERIMENTALS

Materials

Polyethyleneimine derivative containing uracil and homoserine (PEI-Hse-Ura) was prepared by the activated ester method as shown in Scheme 1.[15]   At first, the

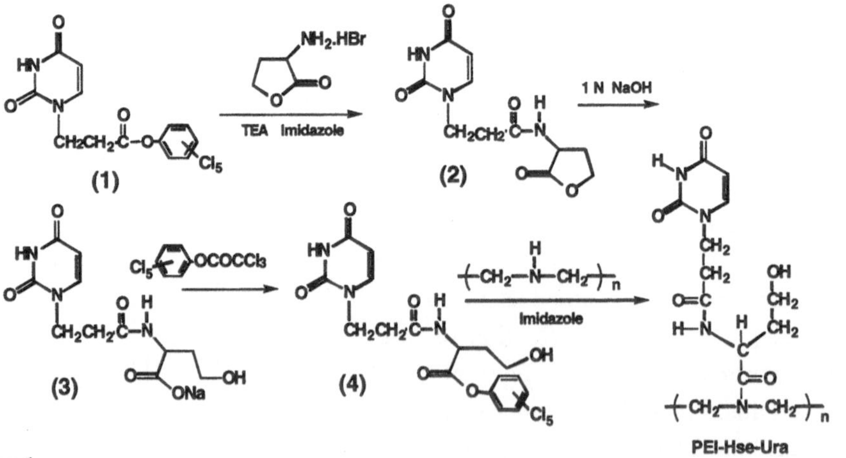

Scheme 1

activated ester of carboxyethyluracil (1) was reacted with (±)-α-amino-γ-butyrolactone hydrobromide, followed by hydrolysis to give the uracil derivative of homoserine (3). The carboxyl group of the obtained compound was activated (4), and was reacted with polyethyleneimine (degree of polymerization was around 500) to give PEI-Hse-Ura. The degree of substitution of this polymer was obtained from UV spectroscopy of hydrolyzed samples to be 94 unit percentage.   The obtained polymer was freely soluble in aqueous buffered solution.

DNA from Calf thymus origin, DNA from Herring sperm, RNA from Yeast, poly (A), and poly (U) were purchased from Yamasa Shoyu Co. Ltd.

Polymer Complex Formation

Interaction studies of these polymers with DNA were carried out using UV spectra.[13-16]   Two kinds of the polymer solutions in Kolthoff buffer at pH 7 (1/10 M KH_2PO_4 - 1/20 M $Na_2B_4O_7 \cdot 10H_2O$) were mixed in various ratios to give a polymer mixture of 10^{-4} M total concentration of nucleic acid base units in solution, and the hypochromicity was measured by UV spectra.   The UV spectra were measured with a JASCO UV-660 spectrometer equipped with a temperature controller at 20 °C.

The circular dichromism (CD) spectra were measured under the same condition used for the UV spectra with a JASCO CD J-40 spectrometer at room temperature.

RESULTS AND DISCUSSION

Interaction with Poly (A) [15]

Figure 4 shows the UV absorbance at 260 nm against various molar ratios of poly (A) to the polyethyleneimine derivative (PEI-Hse-Ura).   The dotted line shows the data for 3 hours after mixing, and the solid line shows the data overnight after mixing.   From the curve in this figure, the maximum hypochromicity value obtained was 54.4 % at 0.5 mole fraction.   Hypochromicity in UV spectra has been widely used to indicate the interaction of nucleic acid derivatives. Therefore, the data indicated that a stable polymer complex was formed between poly (A) with PEI-Hse-Ura by the complementary hydrogen bonding with equimolar nucleic base units (adenine : uracil = 1 : 1).

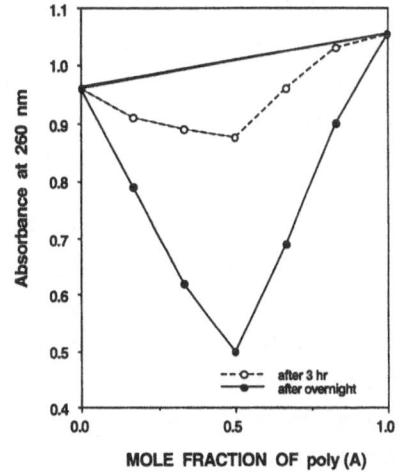

Figure 4. UV Mixing Curves for poly (A) and PEI-Hse-Ura in Kolthoff buffer at pH 7.

To determine whether the interaction between the bases is due to the complementary nucleic acid bases, the interaction of PEI-Hse-Ura with poly (U) was measured under the same condition.   The mixing curve for the poly (U) : PEI-Hse-Ura system in Figure 5 shows no hypochromicity after 3 hours nor even after overnight.   This result indicates that PEI-Hse-Ura did not interact with uracil bases nor the phosphate units in poly (U).   Therefore, the hypochromicity observed for the poly (A) : PEI-Hse-Ura system may be concluded to be caused by the complementary adenine - uracil interaction.

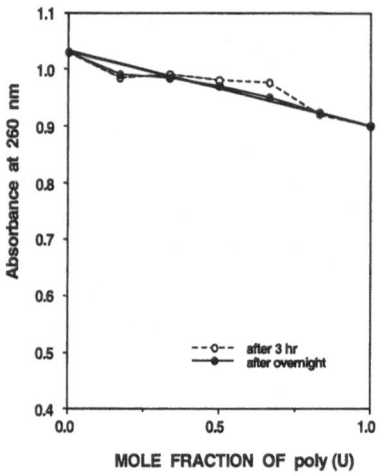

Figure 5. UV Mixing Curves for poly (U) and PEI-Hse-Ura in Kolthoff buffer at pH 7.

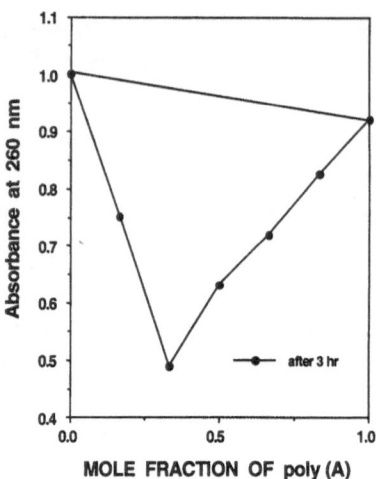

Figure 6. UV Mixing Curves for poly (A) and poly (U).

As a control experiment, the interaction between poly (A) and poly (U) was studied under the same condition used here (Figure 6). The polymer complex formation was observed immediately after mixing of the polymer solutions, and the maximum hypochromicity value was obtained as 40.3 %, which was smaller than the value for the poly (A) : PEI-Hse-Ura system. Compared with the poly (A) : poly (U) system, the formation of the complex was slow for the poly (A) : PEI-Hse-Ura system (Figure 4). This may be due to the intramolecular interaction of uracil bases in PEI-Hse-Ura, that was reported for the polymethacrylate derivative of uracil [17].

Conformation of the Poly (A) : PEI-Hse-Ura Complex

Poly (A) is known to form a right handed single stranded structure caused by stacking of adenine bases in neutral aqueous solution (Figure 7), while the strand is unstable compared with DNA.[18] The conformation of poly (A) was found to be changed by the formation of polymer complex with PEI-Hse-Ura as shown in the CD spectra (Figure 8), where the bands were assigned to poly (A) because PEI-Hse-Ura was optically inactive. The molar ellipticity ([θ]) at 262 nm was plotted against mixing ratio of two polymers in Figure 9. As mentioned above, the hypochromicity in UV spectra indicated that the formation of the complex was slow for the poly (A) : PEI-Hse-Ura system (Figure 4). However, hypochromicity in CD spectra caused by conformational change of poly (A) was observed immediately after mixing of two polymer solutions (Figure 9). From these facts in UV and CD spectra, the

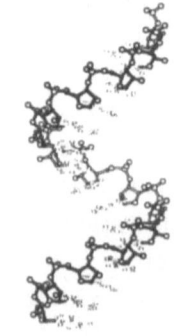

Figure 7. Poly (A).

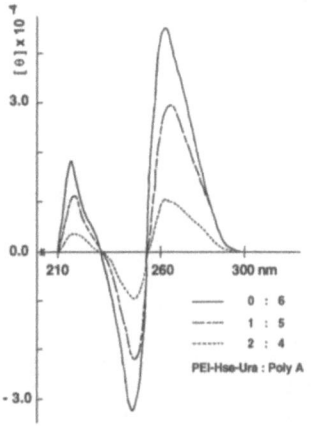

Figure 8. CD Spectra of
poly (A) with PEI-Hse-Ura.

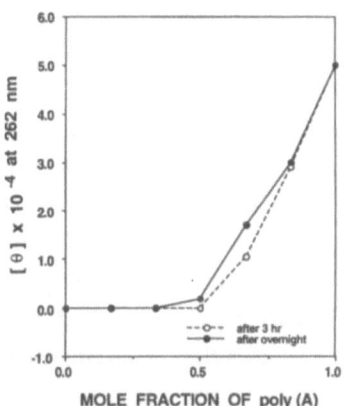

Figure 9. The Value [θ] at 262
nm vs. Mole Fraction of poly (A).

formation of the polymer complex between poly (A) and PEI-Hse-Ura was considered
as follows (Figure 10).

(1) Before mixing, poly (A) forms single stranded structure, and PEI-Hse-Ura exists
in random coiled conformation with intramolecular interaction of uracil bases.

(2) After mixing (3 hours), base pairing between adenine and uracil was partly
formed (UV spectra), and the single stranded structure of poly (A) changed to a random
coiled conformation (CD spectra).

(3) During one night, the intramolecular interaction in PEI-Hse-Ura dissociated
slowly to form the intermolecular base pairing (UV spectra).   The conformation of
poly (A), however, scarcely changed during overnight (CD spectra).

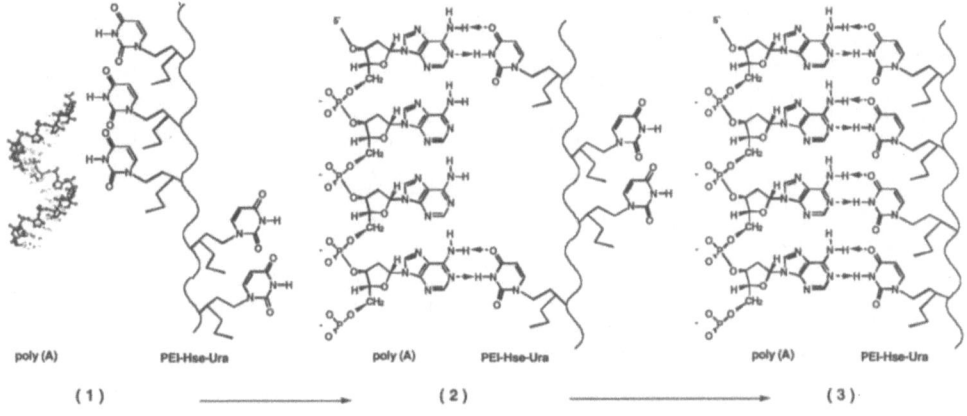

Figure 10.   Formation of the polymer complex between poly (A) and PEI-Hse-Ura.

Interaction with DNA

DNA is known to form a double helical structure by specific hydrogen bondings between complementary nucleic acid bases. Although various structures are known for DNA, the most stable structure of DNA from Calf thymus origins is B-type double stranded conformation (Figure 11).[19] The polyethyleneimine derivative of uracil (PEI-Hse-Ura) was found to form polymer complex with DNA.

Figure 12 shows the UV mixing curve for the DNA : PEI-Hse-Ura system, that suggests the highest hypochromicity value of 47%. The overall stoichiometry of the complex based on the nucleic acid base units was 1 : 2 (DNA : PEI-Hse-Ura), that was different from 1 : 1 for the system poly (A) : PEI-Hse-Ura. From the result and the following facts,

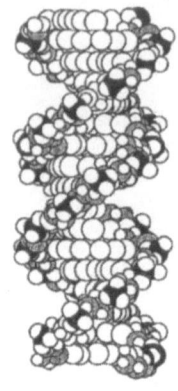

Figure 11. B-Type Calf Thymus DNA

the polymer complex of DNA : PEI-Hse-Ura was concluded as a triple strand (Figure 13). The uracil bases in PEI-Hse-Ura can interact with adenine bases in DNA that already form base pair with thymine in DNA. One adenine base in poly (A) is known to interact with two uracil bases in poly (U) forming poly (A) : (poly (U))$_2$ triple complex.[8]

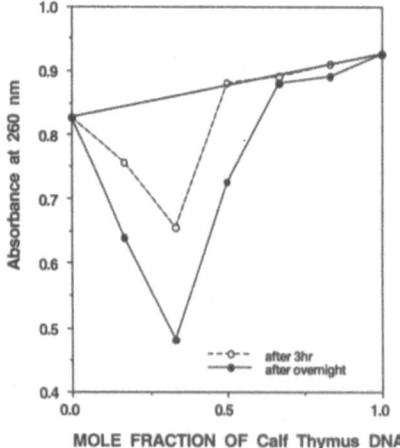

Figure 12. UV Mixing Curves for Calf Thymus DNA and PEI-Hse-Ura.

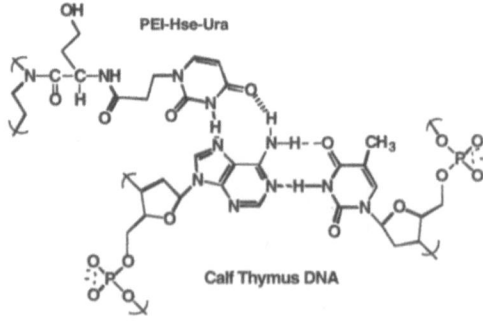

Figure 13. DNA-PEI-Hse-Ura Triple Strand.

PEI-Hse-Ura forms the polymer complex also with DNA from another origin. Figure 14 shows the UV mixing curve for DNA from Herring sperm with PEI-Hse-Ura.

Conformation of the DNA : PEI-Hse-Ura Complex

Drastic change of CD spectra was observed for DNA by addition of the polyethyleneimine derivative as shown in Figure 15. The spectra of DNA (0 : 6 = PEI-Hse-Ura : DNA) shows a typical B type conformation.[20] With increase of PEI-Hse-Ura, the positive band at 280 nm shifted to red (2 : 4 = PEI-Hse-Ura : DNA), and drastic change of spectrum was observed at 3 : 3 molar ratio. The negative band at 280 nm with high intensity indicated a highly condensed form of DNA (Ψ(-) DNA).[21-22] This type conformation may be a super helical structure (Figure 16). Furthermore, with excess PEI-Hse-Ura (5 : 1 = PEI-Hse-Ura : DNA), the spectrum was obtained with negative band at 290 nm and positive band at 245 nm, that should be assign to Z-type conformation of DNA.[23] Simillar change of conformation for DNA was observed for the polyethyleneimine derivatives of thymine having L-serine or D-serine as a spacer (PEI-L-Ser-Thy).

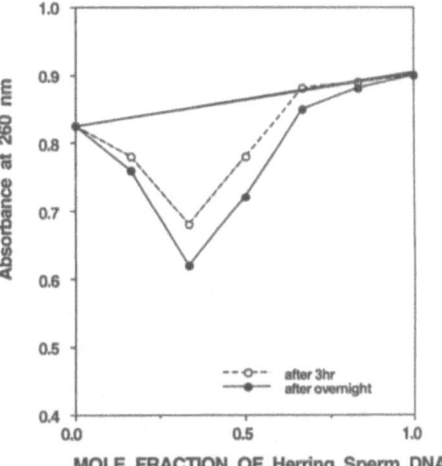

Figure 14. UV Mixing Curves for Herring sperm DNA and PEI-Hse-Ura .

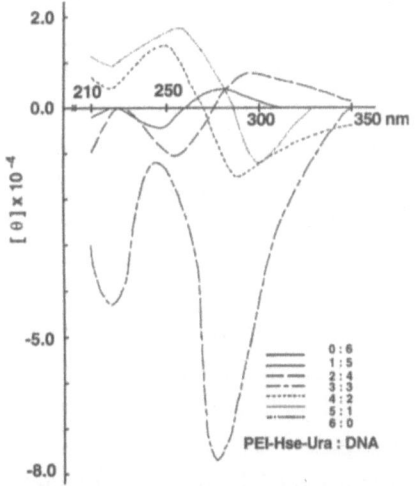

Figure 15. CD Spectra of Calf Thymus DNA with PEI-Hse-Ura.

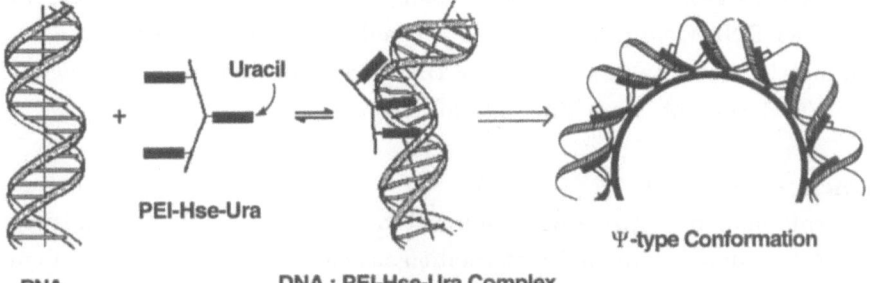

Figure 16. Formation of Polymer Complex and Induction of Super Helix.

Anomalous CD spectrum for the 1 : 1 mixture of DNA with PEI-Hse-Ura was investigated in addition of sodium chloride or metal ions.   Figure 17 shows the spectral change of the 1 : 1 mixture with various sodium chloride concentrations. With increase of NaCl concentration, the intensity of the negative band at 280 nm decreased (up to 100 mM), and became a normal spectrum of DNA with 200 mM of NaCl. The triple strand of DNA : PEI-Hse-Ura probably dissociated to the double strand of DNA, because the double strand of DNA is kown to be stabilized in the presence of neutral salt.[24]

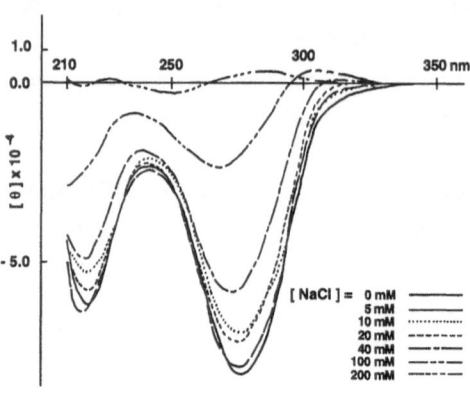

Figure 17. Effect of NaCl on the CD Spectra of Calf Thymus DNA with PEI-Hse-Ura.

Change of the spectra for the 1 : 1 mixture of DNA with PEI-Hse-Ura was also observed in the presence of metal salt.   In the presence of Cu^{2+} (5 x 10^{-5}, and 5 x 10^{-4} M), the intensity of the negative band at 280 nm decreased as shown in Figure 18.   In the presence of calcium ion (Ca^{2+}), however, conformational change was only slightly observed. Metal ion is also kown to interact with the double strand of DNA [25], thus the triple strand of DNA : PEI-Hse-Ura probably dissociated to the copper (II) complex of DNA.

Formation of the polymer complex

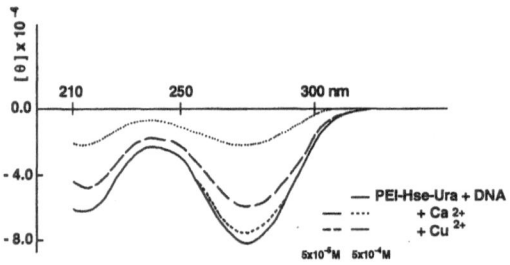

Figure 18. Effect of Metal Ion on the CD Spectra of Calf Thymus DNA with PEI-Hse-Ura.

between DNA from Herring sperm and PEI-Hse-Ura was also observed from the UV spectra (Figure 14).   Conformational change by PEI-Hse-Ura, however, was not observed for DNA from Herring sperm. The double strand of DNA from Herring sperm may be more stable than that of DNA from calf thymus, because guanine - cytosine (G-C) contents were different.[26]

Interaction with RNA

The conformation of RNA is different from that of DNA. RNA from Yeast is reported to form A-type double stranded conformation as shown in Figure 19.[27]   Figure 20 shows the UV mixing curve of RNA with PEI-Hse-Ura. Hypochromicity was observed for this system, but the value was small compared with the system of DNA.   This reason may be caused by the difference of the structure between DNA and RNA. DNA has wide and shallow groove, but RNA has narrow and deep groove where the base pairs are shielded. Uracil bases of PEI-Hse-Ura may penetrate into the major groove of DNA, and forms base pair with adenine of DNA (Figure 16). On the other hand, uracil bases of PEI-Hse-Ura hardly penetrate into the groove of RNA due to the shielded structure of RNA.

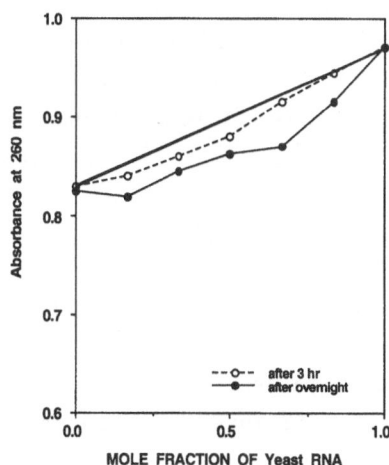

Figure 19. A-Type Yeast RNA.

Figure 20. UV Mixing Curves for Yeast RNA and PEI-Hse-Ura in Kolthoff buffer at pH 7.

CONCLUSION

The polyethylenimine derivatives of uracil having homoserine as a spacer (PEI-Hse-Ura) was soluble in water, and formed polymer complex with poly (A) and DNA by complementary base pairing. PEI-Hse-Ura, however, hardly formed the polymer complex with RNA due to the shielded struture of RNA. PEI-Hse-Ura induced the conformational change of poly (A) and DNA by formation of polymer complex. The induced conformation of DNA with equimolar PEI-Hse-Ura was concluded as a Ψ type conformation.

Polyethyleneimine derivatives of thymine containing L-serine or D-serine as a spacer gave interesting results; the L-serine polymer forms more stable polymer complex with DNA than the D-serine derivative. The detail of the resuls will be reported soon.

REFERENCES

1   K. Takemoto, Y. Inaki (1981), Advances in Polymer Science, Vol. 41, Springer-Verlag, Berlin, Heidelberg, p.1-51

2   K. Takemoto, Y. Inaki (1987), Functional Monomers and Polymers, Marcel Dekker, p. 149-236

3   Y. Inaki, K. Takemoto (1987), Current Topics in Polymer Science, Volume I, Hanser Pub., p. 79-100

4   K. Takemoto, E. Mochizuki, T. Wada, Y. Inaki (1990), Biomimetic Polymers, Plenum Press, New York, p.253-67

5   K. Takemoto, T. Wada, E. Mochizuki, Y. Inaki (1991), Biotechnology and Polymers, Plenum Press, New York, p.31-45

6   Y. Inaki (1992), Progress Polymer Science, vol 17, Pergamon Press, p.515-70

7   C. G. Overberger, Y. Inaki (1979),J. Polymer Sci. Polymer Chem. Ed., 17, 1739~58

8   C. G. Overberger, Y. Inaki, Y. Nambu (1979), J. Polymer Sci. Polymer Chem. Ed., 17, 1759~69

9   Y. Inaki, Y. Sakuma, Y. Suda, K. Takemoto (1982), J. Polymer Sci., Polymer Chem. Ed., 20, 1917~33

10 Y. Sakuma, Y. Inaki, K. Takemoto (1982), J. Polymer Sci., Polymer Chem. Ed., 20, 3431~46

11 Y. Sakuma, Y. Inaki, K. Takemoto (1982), Nucleic Acids Res., Sym. Ser., 11, 269~72

12 Y. Sakuma, Y. Inaki, K. Takemoto (1984), J. Polymer Sci., Polymer Chem. Ed., 22, 2061~82

13 T. Wada, Y. Inaki, K. Takemoto (1988), Polymer J., 20, 1059~68

14 T. Wada, Y. Inaki, K. Takemoto (1989), Polymer J., 21, 11~8

15 T. Wada, Y. Inaki, K. Takemoto (1989), J. Bioact. and Compat. Polym., 4, 25~41

16 T. Wada, E. Mochizuki, Y. Inaki, K. Takemoto (1990), Nucleic Acid. Res., Sym., 22, 113-4

17 S. Fang, Y. Inaki, K. Takemoto (1984), J. Polymer Sci., Polymer Chem. Ed., 22, 2455~67

18 J. Brahms, A. M. Michelson, K. E. Van Holde (1966), J. Mol. Biol., 15, 467~88

19 S. Arnott D. W. L. Hukins (1973), J. Mol. Biol., 81, 93~105

20 V. I. Ivanov, L. E. Minchenkova, A. K. Schyolkina, A. I. Poletayev (1973), Biopolymers, 12, 89~100

21 T. Maniatis, J. H. Venable, L. D. Lerman (1974), J. Mol. Biol., 84, 37~64

22 R. Huey , S. C. Mohr (1981), Biopolymers, 20, 2533~52

23 F. M. Pohl, T. M. Jovin (1972), J. Mol. Biol., 67, 375~96

24 J. Marmur, P. Doty(1962), J. Mol. Biol., 5, 109~18

25 G. L. Eichhorn, Y. A. Shin (1968), J. Am. Chem. Soc., 90, 7323~8

26 C. Bostock (1980), Trends Biochem. Sci., 5, 117~9

27 S. Arnott, D. W. L. Hukins, S. D. Dover, W. Fuller, A. R. Hodgson (1973), J. Mol. Biol., 81, 107~22

Helical Polymers:
Synthesis, Conformation, and Function

Yoshio Okamoto, Tamaki Nakano, and Shigeki Habaue

Department of Applied Chemistry, School of Engineering
Nagoya University, Furo-cho, Chikusa-ku, Nagoya 464-01, Japan

Abstract: Asymmetric polymerization of acrylic monomers and an isocyanate derivative leading to a helical conformation is described. 1-Phenyldibenzosuberyl methacrylate, triphenylmethyl acrylate, and N,N-diphenylacrylamide were polymerized using the complexes of diphenylethylenediamine monolithium amide with (-)-sparteine, (S,S)-(+)-2,3-dimethoxy-1,4-bis(dimethylamino)butane, and (S)-(+)-1-(2-pyrrolidinylmethyl)pyrrolidine. The methacrylate gave a purely one-handed helical, optically active polymer ($[\alpha]_{365}$ +1670° ~ +1780°) having a perfect isotactic structure; the polymer exhibited chiral recognition ability to some racemic compounds. This monomer gave optically active polymers also by radical mechanism. The acrylate also gave an optically active, helical polymer ($[\alpha]_{365}$ +102°) having a dyad isotacticity of 70% using the pyrrolidine complex. The poly(methyl acrylate) derived from the polyacrylate showed a specific rotation which appears to be based on configurational chirality of the main chain. The acrylamide gave an optically active polymer ($[\alpha]_{365}$ -429°) which may have a helical structure by asymmetric anionic polymerization; the polymer had a dyad isotacticity of 86%. m-Methylphenyl isocyanate was polymerized using optically active lithium alkoxides and amides to produce a polymer having a chiral group at the α-end. A prevailing one-handed helical structure was induced for the main chain due to the chiral end group.

INTRODUCTION

Many stereoregular polymers have a helical conformation in the solid state and some of them can maintain the structure in solution. Helical structure is chiral and therefore one-handed helical polymer can be optically active [1]. Helical polymers are interesting because they may show chiral recognition for enantiomers. Several helical polymers including polymethacrylates and polysaccharide derivatives have been practically used as stationary phases for HPLC to separate a wide range of enantiomers [1, 2, 3]. Optically active polymers having a prevailing one-handed helicity have been prepared

M. Kamachi · A. Nakamura (Eds)
New Macromolecular Architecture and Functions
Proceedings of the OUMS '95 Toyonaka, Osaka, Japan, 2-5 June, 1995
© Springer-Verlag Berlin Heidelberg 1996

through asymmetric (helix-sense-selective) polymerization of prochiral monomers such as methacrylates, acrylates, *N,N*-disubstituted acrylamides, and isocyanates by using optically active anionic initiators. Here, helix-sense-selective radical and anionic polymerizations of 1-phenyldibenzosuberyl methacrylate (PDBSMA), triphenylmethyl acrylate (TrA), *N,N*-diphenylacrylamide (DPAA), and *m*-methylphenyl isocyanate (*m*-MPI) are mainly discussed.

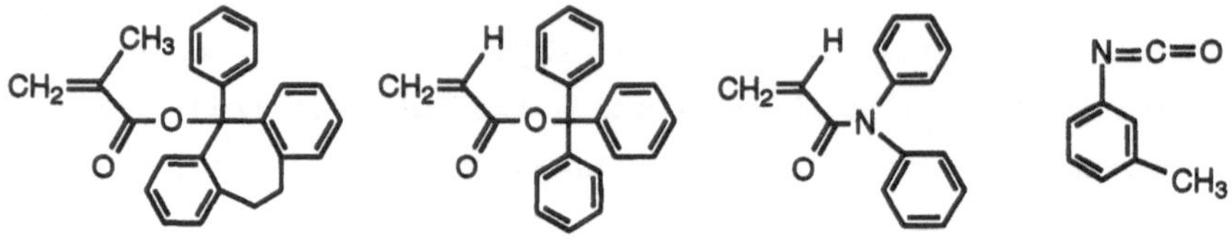

PDBSMA   TrA   DPAA   *m*-MPI

1-PHENYLDIBENZOSUBERYL METHACRYLATE (PDBSMA) [4, 5]

Anionic Polymerization

One-handed helical poly(triphenylmethyl methacrylate) [poly(TrMA)] is the first example of such kind of a vinyl polymer [6]. The optically active poly(TrMA) can resolve many classes of racemic compounds when used as a stationary phase for HPLC; however, the ester linkage in the poly(TrMA) is readily solvolyzed by methanol, an effective HPLC solvent, and the resolution ability is gradually lost. PDBSMA was designed after TrMA in order to improve this shortcoming of poly(TrMA). We assumed that tying two phenyl groups of TrMA should reduce the planarity of the tertiary cation which is involved in the solvolysis reaction and therefore retards the solvolysis. PDBSMA monomer had a rate constant of 0.466 hr^{-1} for the reaction in **Scheme 1** which is ca. six times as low as the rate constant for TrMA monomer, confirming the rationality of our monomer design.

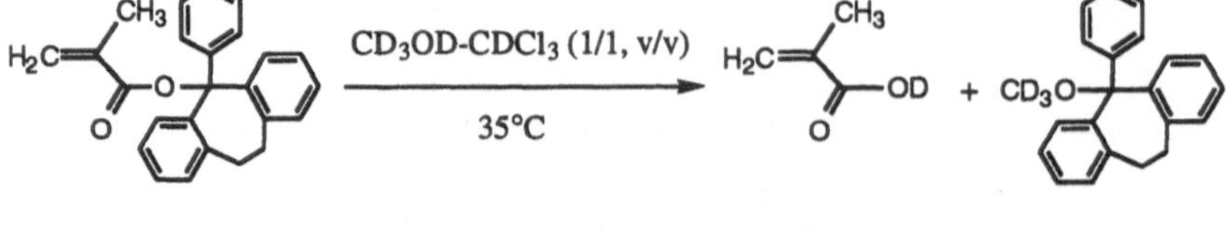

Scheme 1. Methanolysis of PDBSMA

An almost perfectly isotactic polymer is obtainable from PDBSMA by anionic polymerization using the complexes of diphenylethylenediamine monolithium amide (DPEDA-Li) with (-)-sparteine ((-)-Sp), (*S,S*)-(+)-2,3-dimethoxy-1,4-

bis(dimethylamino)butane ((+)-DDB), and (S)-(+)-1-(2-pyrrolidinylmethyl)pyrrolidine ((+)-PMP) (**Table 1**). The polymers show high optical activity and intense circular dichroism absorption bands. GPC analysis of the polymers using a UV and a polarimetric detectors showed that the chromatograms from the two detectors had a similar shape. Furthermore, chiral HPLC analysis of the (+)-polymers revealed no clear existence of (-)-fraction. These observations clearly indicate that the polymers have a helical conformation with exclusively right- or left-handed helicity. The optically active polymer showed chiral recognition ability to several racemates though the ability was lower than that of poly(TrMA).

Table 1. Asymmetric anionic polymerization of PDBSMA using DPEDA-Li complexes with chiral ligands in toluene at -78°C[a]

Run	Chiral Ligand	Yield[b]	DP	$[\alpha]_{365}$[c]
1	(+)-DDB	86%	48	+1778°
2	(+)-PMP	95%	43	+1755°
3	(-)-Sp	71%	80	+1670°

[a]Conditions: [PDBSMA]/[Li] = 20; time 24 hr. Monomer conversion was quantitative in all the runs.
[b]Benzene-hexane (1/1)-insoluble (oligomer-free) part.
[c]In CHCl$_3$.

Radical Polymerization

PDBSMA gives a highly isotactic polymer also by radical polymerization; the triad isotacticity (*mm*) is higher than 98% [4]. This makes it possible to synthesize optically active poly(PDBSMA) with a single handed helicity by radical mechanism using chiral additives (helix-sense-selective radical polymerization).

Radical polymerization of PDBSMA in a mixture of menthol and toluene gave an optically active, highly isotactic polymer (**Table 2**). The polymers prepared in the presence of (+)- and (-)-isomers of menthol showed specific rotation of opposite sign. In addition, the spectral pattern of circular dichroism (CD) of the polymer obtained using (+)-menthol was similar to that of the anionically obtained polymer and almost the

92

mirror image of that of the polymer obtained using (-)-menthol. These results indicate that the chiroptical properties of the polymers are based on a helical conformation with right- or left-handed helicity in excess. The chiral induction observed in the radical polymerization may be based either on difference in propagation rate of right- and left-handed radicals in the chiral medium or on chain transfer mechanism.

Table 2. Radical polymerization of PDBSMA in a mixture of toluene and menthol[a]

			THF-soluble, B/H-insoluble part[c]		
Run	Reaction medium	Yield[b]	Yield	DP	$[\alpha]_{365}$[d]
1	(+)-menthol-toluene	43%	1%	50	-200°
2	(-)-menthol-toluene	45%	1%	50	+180°

[a]Conditions: Monomer 0.1g, toluene 1.5 ml, menthol 4 g.
[b]Hexane-insoluble part of the product.
[c]The product was fractionated first by THF and then by benzene-hexane (1/1) (B/H) to remove high-molecular-weight fraction and oligomers, respectively.
[d]In THF.

Polymerization in the presence of (-)-menthanethiol and (+)- and (-)-neomenthanethiol also gave optically active polymers whose specific rotation ($[\alpha]_{365}$) ranges from 60° to 140° depending on the structure and concentration of the thiol used in the polymerization. In this case, chiral induction can take place through selective termination of right- or left-handed helical radical by the chiral thiol (hydrogen transfer) or initiation reaction by the chiral thio radical formed by the hydrogen transfer process. Radical copolymerization of PDBSMA with a small amount of optically active (-)-PPyoTMA was also effective in obtaining an optically active polymer having a single handed helical conformation in excess.

TRIPHENYLMETHYL ACRYLATE (TrA) [7]

In the case of an α,α-disubstituted vinyl monomer such as bulky methacrylates, the α-methyl group in addition to a bulky ester group also plays an important role in forming the chiral conformation in the asymmetric (helix-sense-selective) polymerization. In general, the control of stereoregularity in the polymerization of acrylates is more difficult than that of methacrylates. For example, it has been known that triphenylmethyl acrylate (TrA) forms a polymer with much less isotacticity in anionic polymerization as well as in radical polymerization in comparison with TrMA [8].

The polymerization of TrA was performed in a manner similar to that in the polymerization of PDBSMA using the complexes of DPEDA-Li with chiral ligands, (–)-Sp, (+)-DDB, and (+)-PMP in toluene at –78°C for 24 h. The results of polymerization are summarized in **Table 3**. The polymerization proceeded quantitatively in all cases, and the tacticity and the specific rotation of polymers were greatly affected by the ligands. The specific rotation of the polymers seems to increase with an increase of the isotacticity, and the (+)-PMP system afforded the polymer of the highest specific rotation ($[\alpha]_{365}^{25}$ +102°) and isotacticity (m = 70%) (Run 1). However, these values are much smaller than those ($[\alpha]_{365}^{25}$ +1500°, mm > 99%) of one-handed helical, optically active poly(TrMA) [6].

Table 3. Polymerization of TrA by DPEDA-Li complexes in toluene at -78°C for 24h[a]

Run	Chiral ligand	Yield (%)[b]	DP[c]	Mw / Mn[c]	Tacticity, m / r	$[\alpha]_{365}$[d]
1	(+)-PMP	96	46	1.28	70 / 30	+102°[e]
2	(–)-Sp	98	57	2.94	64 / 36	n.d.[f]
3	(+)-DDB	100	61	1.38	49 / 51	–7°

[a][TrA] / [Initiator] = 20. [b]Methanol-insoluble part. [c]Determined by GPC (polystyrene standard). [d]Measured in CHCl3 at 25°C. [e]$[\alpha]_D$ +22° (c 1.53). [f]Not determined. The obtained polymer was partly insoluble in common organic solvents.

The CD spectrum of (+)-poly(TrA) (Run 1 in **Table 3**) is shown in **Figure 1** (a). The spectrum demonstrates the positive peaks at 210 and 230 nm which may be ascribed to the absorption due to the aromatic and carbonyl groups, respectively. This spectral pattern is quite similar to that of the one-handed helical, optically active poly(TrMA). This suggests that the optical activity of the poly(TrA) may be attributed to a partially one-handed helical structure of the polymer chain. The degree of one-handedness must be lower than that of the poly(TrMA) judging from the smaller optical activity and isotacticity of the polymer.

The specific rotation ($[\alpha]_{365}^{25}$ +85° (c 0.91)) of (+)-poly(TrA) in THF did not change in 50 min at 25°C. However, the optical activity gradually increased at 60°C and reached about +138° after 20 min. Then the solution became turbid. These results indicate that the (+)-poly(TrA) is conformationally stable at room temperature but some conformational change and association occur at a higher temperature. In chloroform, the (+)-poly(TrA) precipitated immediately after the dissolution.

To obtain information about the contribution of a configurational factor to the chiroptical property of the polymer, the (+)-poly(TrA) (Run 1 in **Table 3**) was converted to poly(methyl acrylate) (PMA) by hydrolysis in methanol containing a small amount of hydrochloric acid followed by methylation with diazomethane. The PMA showed a small negative specific rotation ($[\alpha]_{365}^{25}$ –18°), which is opposite in sign to that of the (+)-poly(TrA). The CD spectrum of the PMA was quite different in pattern from that for poly(TrA) (**Figure 1**). The optical activity of the PMA may be mainly due to the chirality near polymer ends because the CD peaks of PMA are attributed mainly to the absorption of the initiator residue, an *N,N'*-diphenylethylenediamino group. The induction of the configurational asymmetry clearly occurs at the initial stage of the polymerization and may continue through the polymerization to produce the prevailing conformational asymmetry in the polymer chain.

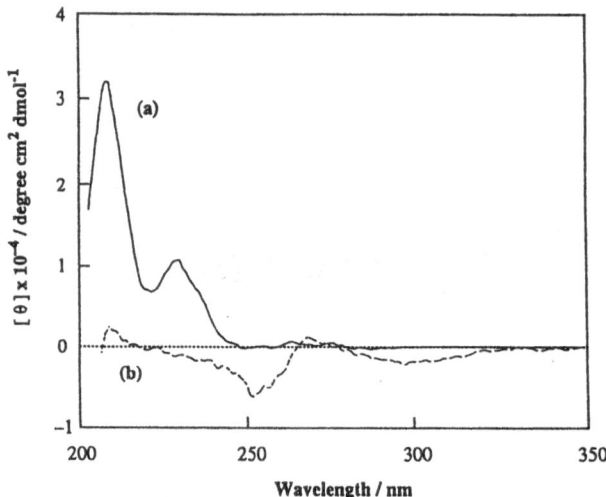

Figure 1. CD spectra of (+)-poly(TrA) (a) and (–)-PMA (b) derived from the (+)-poly(TrA) in THF. The molar concentration of (+)-poly(TrA) was calculated on the basis of the monomeric unit (Mw = 314) and that of (–)-PMA on the basis of the polymer molecules (Mw = 4.9 x 10^3).

N,N-DIPHENYLACRYLAMIDE (DPAA) [9, 10]

The N,N-diphenylamino group in DPAA is smaller than triphenylmethyl group in TrA, but the former is in the β-position with respect to the C=C double bond while the latter is in the γ-position. Therefore, there is a possibility to form a chiral polymer with helical conformation in the asymmetric polymerization of DPAA. The asymmetric anionic polymerization of DPAA was carried out using the complexes of fluorenyllithium (FlLi) with chiral ligands, (–)-Sp and (+)-PMP in toluene at –98°C. The tacticity of the obtained polymer was determined by ^{1}H NMR analysis on poly(methyl acrylate) (PMA) [11, 12], which was carefully converted from poly(DPAA) by solvolysis in a mixture of sulfuric acid and methanol followed by methylation with diazomethane.

Table 4 shows the results of the polymerization of DPAA. The polymerization was proceeded smoothly to afford poly(DPAA) almost quantitatively (Run 1). Some characteristic features of the polymerization are as follows: (1) The fact that methanol-insoluble polymer was obtained even under the monomer to initiator ratios one or two indicates that the propagation rate is much faster than the initiation (Run 3 and 4). (2) The (–)-Sp-FlLi system gave the polymers possessing the high optical rotation ($[\alpha]_{365}^{25}$ –260°) and isotacticity (m = 71%), whereas (+)-PMP was ineffective as a chiral ligand in the polymerization of DPAA (Run 5). But the specific rotation of poly(DPAA) was much smaller than that of one-handed helical, optically active poly(TrMA) obtained under the same reaction conditions and opposite in sign. (3) The specific rotation of the obtained polymers was increased with an increase of the isotacticity, and the value reached to –429° ($[\alpha]_{365}^{25}$) when the isotacticity (m) was 86% (Run 4).

Table 4. Polymerization of DPAA with (–)-Sp-FlLi in toluene at –98°C

Run	[M] / [I]	Time	Yield (%)[a]	Mn[b]	Mw / Mn[b]	Tacticity, m / r	$[\alpha]_{365}$[c]
1	20	1h	95	4600	1.27	71 / 29	–260°
2[d]	20	1h	91	4200	1.16	64 / 36	–201°
3	1	1h	26	3000	1.10	88 / 12	–418°
4	2	1h	32	3000	1.12	86 / 14	–429° [e]
5[f]	20	2h	90	6900	1.22	44 / 56	–3° [g]

[a]Methanol-insoluble part. [b]Determined by GPC (polystyrene standard).
[c]In CHCl$_3$-CF$_3$CO$_2$H (c 0.5, 25°C). [d]Polymerization temp.: –78°C.
[e]$[\alpha]_D$ –65°. [f]Polymerization was carried out using the (+)-PMP-FlLi system.
[g]In THF (c 1.0, 25°C).

The CD spectrum of (–)-poly(DPAA) (Run 3 in **Table 4**) is depicted in **Figure 2**. The broad negative peaks in the spectrum indicate that the phenyl groups of the polymer exist under chiral circumstance. The optical activity may be attributed to a chiral conformation of the polymer chain produced through the polymerization process.

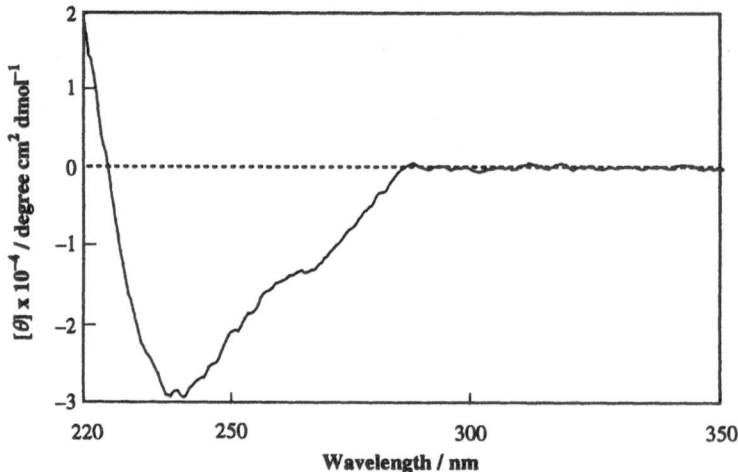

Figure 2. CD spectrum of (–)-poly(DPAA) (Run 3 in **Table 4**, $[\alpha]_{365}^{25}$ –418°) in chloroform / CF_3CO_2H (36 / 1, v / v).

POLY(*m*-METHYLPHENYL ISOCYANATE) (*m*-MPI) [13,14]

Polyisocyanates are known to have a helical main chain. The helical conformation is dynamic and it consists of equal amounts of right-handed and left-handed parts separated by helix reversal points moving along the main chain; therefore, poly(alkyl isocyanate)s baring no chiral side group is optically inactive. However, it is possible to obtain an optically active polymer having an excessive right- or left-handed helical content from achiral isocyanates by polymerization using an optically active anionic initiator such as shown below. The obtained polymer has a chiral α-end group and a certain length of the main chain staring from the α-end takes a one-handed helical conformation by the influence of the chiral group. Therefore, a polymer with higher molecular weight has a smaller optical activity.

Li-menthoxide   Li-borneoxide   Li-Chirald

Li-DGG   Li-PMP   Li-MMP

The results of polymerization *m*-methylphenyl isocyanate (*m*-MPI) are shown in **Table 5** [13]. The polymer obtained using Li-MMP showed a high dextrorotation, while the unimer of m-MPI bearing an MMP residue was levorotatory ($[\alpha]_{365}$ -443°). Additionally, the polymers contain no asymmetric centers in the main chain. These results strongly suggest that the optical activity of the polymers is based on the prevailing helicity of the polymer chain. Poly(aromatic isocyanate)s must have some rigidity on the main chain which was not expected previously [15].

Table 5. Polymerization of *m*-MPI in tetrahydrofuran at -98°C

Run	Initiator	$[\alpha]_{365}{}^{a}$ (deg)	*Mw* (x 10^4)	*Mn* (x 10^4)	*Mw/Mn*
1	Li-Chirald	+8.9	6.5	4.6	1.4
2	Li-borneoxide	+2.9	5.2	3.8	1.4
3	Li-menthoxide	-2.4	7.1	5.4	1.3
4	Li-DGG	-67	4.9	3.4	1.4
5	Li-PMP	+173	1.4	1.1	1.2
6	Li-MMP	+819	1.1	0.9	1.2

[a]In tetrahydrofuran.

m-MPI oligomer
(n = 1: unimer)

In order to obtain information on the helical structure of poly(m-MPI), uniform oligomers of m-MPI having the degree of polymerization (DP) of 1-18 were isolated using HPLC [14]. The specific rotation ($[\alpha]_{365}$) increased from ca. +1100° to ca. +2000° with an increase in DP in the range of $3 \leq DP \leq 12$ and then gradually decreased in the DP range above 12. These results mean that the oligomers having DP of 12 or smaller have a helical structure without a helix reversal point and optical activity increases as the helical structure persists longer with an increase in DP in the DP range of 3 to 12. The oligomers having DP of 13 or larger may start to have helix reversals. In higher DP ranges (polymers), above DP = 70, the specific rotation was found to be inversely proportional to DP of the polymer where the data fitted to the equation (1):

$$[\alpha] = k/DP \qquad (1)$$

However, in the DP range smaller than 30, the specific rotation could not be expressed by eq. (1). This may mean that the influence of the chiral α-end group completely disappears at DP = 30-70.

REFERENCES

1   Okamoto Y, Nakano T (1994) Chem Rev 94: 349
2   Okamoto Y, Hatada K (1986) J Liq Chromatogr 9: 396
3   Okamoto Y, Kaida Y (1994) J Chromatogr A 666:1994
4   Nakano T, Mori M, Okamoto Y (1993) Macromolecules 26: 867
5   Okamoto Y, Nakano T, Shikisai Y, Mori M (1995) Macromol Symp 89: 479
6   Nakano T, Okamoto Y, Hatada K (1992) J Am Chem Soc 114: 1318; Okamoto Y, Suzuki K, Ohta K, Hatada K, Yuki H (1979) J Am Chem Soc 101: 4763
7   Habaue S, Tanaka T, Okamoto Y (1995) Macromolecules in press
8   Matsuzaki K, Uryu T, Kanai T, Hosonuma K, Matsubara T, Tachikawa H, Yamada M, Okuzono S (1977) Makromol Chem 178:11
9   Okamoto Y, Adachi M, Shohi H, Yuki H (1981) Polym J 13:175
10 Okamoto Y, Hayashida H, Hatada K (1989) Polym J 21:543
11 Matsuzaki K, Uryu T, Ishida (1967) J Polym Sci, Part A 5:2167
12 Suzuki T, Santee ER Jr, Harwood HJ, Vogl O, Tanaka T (1974) J Polym Sci, Polym Lett Ed 12:635
13 Okamoto Y, Matsuda M, Nakano T, Yashima E (1994) J Polym Sci, Part A, Polym Chem 32: 309
14 Maeda K, Matsuda M, Nakano T, Okamoto Y (1995) Polym J 27: 141
15 Bur AJ, Fetters LJ (1976) Chem Rev 76: 727

The Dendritic Box; Synthesis, Properties, and Applications

J.F.G.A. Jansen[1], E.M.M. de Brabander - van den Berg[2], E.W. Meijer[1*]

1) Laboratory of Organic Chemistry, Eindhoven University of Technology, P.O. Box 513, 5600 MB Eindhoven, The Netherlands and 2) DSM Research, P.O. Box 18, 6160 MD Geleen, The Netherlands

Abstract: A dendritic structure with a densely-packed shell is prepared via modification of the fifth generation poly(propylene imine) dendrimers with N-BOC-L-phenylalanine groups. This so-called dendritic box has a diameter of approximately 4.5 nm, as determined by DLS and SAXS, and has internal cavities which can be used for the encapsulation of guest molecules. Supramolecular encapsulation is achieved by performing the construction of the box in the presence of the guest molecules, followed by dialysis to remove excess and adhered guest. A number of properties of encapsulated guests is critically influenced by the host, like the observation of induced optically activity and a solvent-independent fluorescence of dyes encapsulated as well as the formation of a triplet radical pair of 3-carboxy-proxyl. A shape-selective liberation of encapsulated guests is accomplished by a two-step hydrolysis of the shell. The potentials of these boxes as new supramolecular architectures within the nanometer regime are discussed .

INTRODUCTION

The interest in dendrimeric macromolecules arises from the unique properties of these highly branched structures that have a defined number of generations and functional end groups [1,2]. The high degree of control over molecular weight and shape has led to the synthesis of unimolecular micelles [3], spherical and cone-shape mesostructures [4], as well as stratified dendrimers possessing generations of different structure [5]. Diameters of the spherical dendrimers range from 3-10 nanometers, enabling these structures to be building blocks of a new chemistry set [6].

After the initial reports on cascade molecules [7], proposals have been made for the construction and applications of guest-host systems made out of dendrimers [1,2,8]. The concept of topological trapping by core-shell molecules is based on the fact that, at some stage in the synthesis of dendrimers, the space available for the new generation or end-group modification is not sufficient to accommodate all the atoms required for complete conversion (the so-called sterically induced stoichiometry) [1]. We will discuss here the synthesis of a dendritic box consisting of a flexible core with a rigid shell [9]. These boxes have internal cavities available in which guest molecules can be physically entrapped due to the rigid shell.

THE DENDRITIC BOX

The flexible core of our dendritic box is based on poly(propylene imine) dendrimers which were synthesized by the divergent approach [10]. A repetitive reaction sequence using the double Michael addition of a primary amine to acrylonitrile followed by the heterogeneously catalyzed hydrogenation of the nitriles to primary amines, yields diaminobutane-based poly(propylene imine) dendrimers DAB-*dendr*-$(NH_2)_n$ with n= 4, 8, 16, 32, 64, and 128 primary amine end groups. The unmodified dendrimers are very flexible and possess glass transition temperatures of approximately -40 °C and -65 °C for the CN- and NH_2-terminated dendrimers, respectively [10]. More recently a variety of end-group modifications have been reported [11]. For the construction of the rigid shell of the dendritic box a critical end-group modification of the cascade polyamine with an appropriate

M. Kamachi · A. Nakamura (Eds)
New Macromolecular Architecture and Functions
Proceedings of the OUMS '95 Toyonaka, Osaka, Japan, 2-5 June, 1995
© Springer-Verlag Berlin Heidelberg 1996

bulky group is performed. For instance, the N-hydroxy-succinimide ester of a *tert*-butyloxycarbonyl (t-BOC)-protected L-phenylalanine is brought into reaction with the fifth generation poly(propylene imine) dendrimer in a CH₂Cl₂-triethylamine mixture (figure 1). Extended washing procedures were used to obtain pure dendritic box with a molecular weights of almost 24,000 [9].

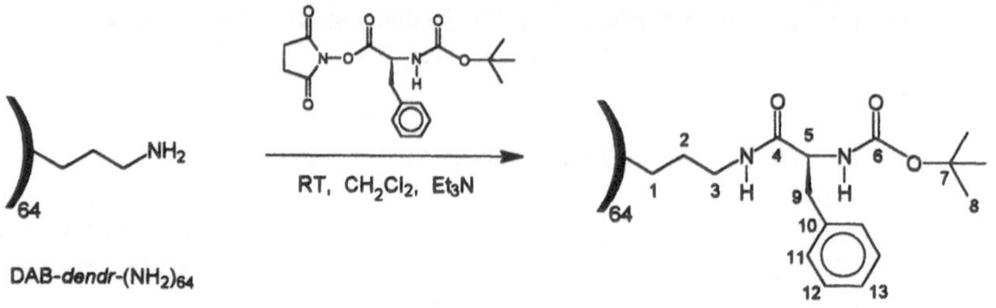

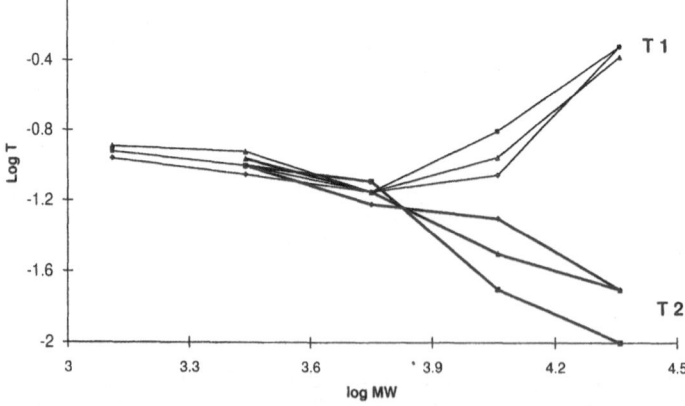

Figure 1: Schematic presentation of the synthesis of the amino acid-terminated poly(propylene imine) dendrimers, including an atomic numbering of the shell of the dendritic box.

Structure elucidation of the dendritic box was performed with all possible characterization techniques; IR, UV, ^{1}H- and ^{13}C-NMR spectroscopy data are all in agreement with the structure assigned. However, the resonances in the ^{13}C-NMR spectra showed a significant line broadening for the higher generations. Spin-lattice (T_1) and spin-spin (T_2) relaxation measurements were performed and the results for the shell atoms were compared with the corresponding data of the other lower generations (figure 2). The observed increase of T_1 relaxation times after the third generation is indicative of a decrease in molecular motion for the higher generations; an almost solid-phase behavior of the shell in solution is proposed. Further evidence for this close packing of the shell is found from chiroptical studies (see later). Presumably, intramolecular hydrogen bonding between several L-Phe residues in the shell is contributing to this solid-phase character. Unfortunately, MALDI-TOF and electrospray mass spectrometry studies have not been successful yet.

Figure 2: Double logarithmic graph of relaxation data (carbon T_1 and T_2) versus molecular weight (generation) for the carbon atoms 1, 2, and 3 (see figure 1) as recorded at 75 Mhz in chloroform.

CHARMm molecular mechanics calculations of the DAB-*dendr*-(N-t-BOC-L-Phe)₆₄ are performed to get insight into the three-dimensional structure. A globularly architecture is found with an estimated radius of 23 ± 3 Å. Dynamic light scattering studies of the dendritic box in solution showed single particle behavior with a radius of gyration of 17 ± 4 Å (which resembles a radius of the box of 22 Å). Finally, SAXS measurements gave a radius of gyration of 18 Å for the dendritic box.

The choice of L-Phe as the amino acid component of the shell has been made from a study in which we compared a variety of amino acids of different size. By using larger amino acids like L-Trp it is not possible anymore to modify all the end groups due to the restricted space available. On the other hand by performing the modification reaction with smaller amino acids, like L-Ala and L-Leu such a dense packing is not achieved, as concluded from NMR and modeling studies, as well as the encapsulation experiments described in the next paragraph. L-Tyr in the t-BOC protected form is comparable with L-Phe as shell component with respect to dense packing, but lacks the good solubility in most organic solvents. Therefore, we selected the DAB-*dendr*-(NH-t-BOC-L-Phe)$_{64}$ as the dendritic box, being a nanometer-sized host system for a variety of guest molecules [9].

ENCAPSULATION OF GUEST MOLECULES INTO THE DENDRITIC BOX

The experimental and modeling results prompted us to propose that we prepared molecules with a solid shell and a flexible core that will have internal cavities available for guest molecules. As the shell is constructed in the last step, it is possible to perform this coupling reaction in the presence of guest molecules. In fact, we encapsulated molecules with some affinity for tertiary amines within the dendritic box. Excess of guest and/or traces of guests adhering to the surface are removed by extensive washing and/or dialysis. When a dendrimer of lower generation was used, the shell is not dense enough to capture the guests and they were removed by extraction. A large variety of guest molecules have been encapsulated and this opens a plethora of interesting chemical and biochemical applications. We will discuss some of these nanometer-sized guest-host systems here as well as the properties of the guest molecules that are so critically influenced by the dendritic box.

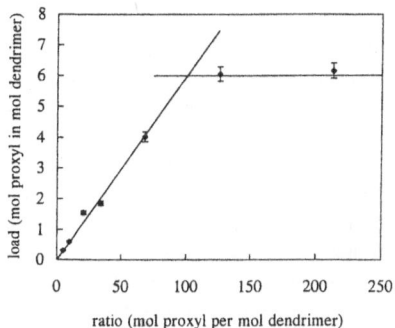

Figure 3: Number of 3-carboxy-proxyl radicals trapped in the dendritic box as determined by ESR spectroscopy versus the molar ratio of radical and dendrimer in the initial solution prior to the encapsulation reaction.

By carring out the encapsulation reaction in the presence of a varying concentration of 3-carboxy-proxyl, the number of entrapped radicals could be varied from 0.3 to 6.0 molecules per dendritic box as determined by ESR spectroscopy [12]. The number of 3-carboxy-proxyl radicals in the dendritic box does not increase above 6 (figure 3), clearly demonstrating that the maximum attainable number of radicals is restricted by the shape of the cavities of the box. The ESR spectra of 3-carboxy-proxyl@DAB-*dendr*-(NH-t-BOC-LPhe)$_{64}$ dissolved in 2-methyltetrahydrofuran are strongly temperature dependent. At 305 K an essential isotropic ^{14}N-coupled ESR spectrum is observed, characteristic for a rapid rotational diffusion of the radical spin probes. Lowering the temperature results in a decreasing intensity of the isotropic spectrum and the appearance of an anisotropic ESR spectrum, consistent with a more restricted motion of the spin probe. In the temperature range from 150 to 250 K a superposition of the motionally narrowed (isotropic with

$A_{iso}(N)$=1.40-1.42 mT) and the slowmotion (anisotropic with A_{zz} (N)=3.38 mT) spectrum is observed. This superposition indicates that the micro-environment of the encapsulated 3-carboxyl-proxyl molecules is not uniform over the interior of the dendritic box. A solid sample of 3-carboxy-proxyl@DAB-*dendr*-(NH-t-BOC-LPhe)$_{64}$ with more than 1.6 molecules per box shows at lower temperatures a (partial) ferromagnetic alignment of the radicals. The observation of a Δ_{ms} = 2 ESR transition exhibiting a partially resolved 1:2:3:2:1 hyperfine coupling pattern due to two ^{14}N nuclei with A(pair) =1/2A(3-carboxy-proxyl) showed unambiguous spectral evidence of the presence of a triplet-state readical pair (figure 4). The intensity of the Δ_{ms} = 2 signal follows Curie law (I =C/T) between 4.2 and 100 K, consistent with a triplet ground state. To the best of our knowledge this is the first observation of an intermolecular ferromagnetic exchange interaction in a non-crystalline guest-host assembly. Since these type of interactions are often observed intramolecularly or in organic crystals, we are prompted to conclude that the dendritic box possesses some peculiar ordering properties apparently dictated by the architecture of the dendritic skeleton.

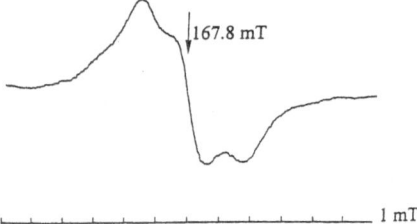

Figure 4: Δ_{ms} = 2 ESR spectrum of a solid sample of 3-carboxy-proxyl@DAB-*dendr*-(NH-t-BOC-LPhe)$_{64}$ at 4.2 K.

As another example we have encapsulated a variety of organic dye molecules into the dendritic box [9]. Bengal Rose is encapsulated in a similar fashion as the spin probe described above. The number of Bengal Rose molecules encapsulated could be estimated after prolonged dialysis by comparison of the UV spectra of guests that are inside or outside of the box. The relation between the number of encapsulated molecules of Bengal Rose as a function of the concentration of Bengal Rose used in the shell-forming reaction is depicted in figure 5. Also in this case the maximum number of guest molecules attainable is limited to in this case 4. It is tempting to propose that each of the 4 guest molecules is occupying one large cavity present in the dendritic box. Although the absorption spectra of Bengal Rose and Bengal Rose@DAB-*dendr*-(NH-t-BOC-LPhe)$_{64}$ are identical, there is a large difference in the fluorescence spectra as recorded in CHCl$_3$.The strong fluorescence at λ_{max} = 600 nm for Bengal Rose@DAB-*dendr*-(NH-t-BOC-LPhe)$_{64}$ is completely absent in the case of the supramolecular isomer of Bengal Rose out of the box. In the latter the fluorescence is quenched effectively. The emission of the guest-host system is relatively insensitive to solvent effects, hence, we believe that we have prepared a fluorescent sphere with an environmental-independent emission profile.

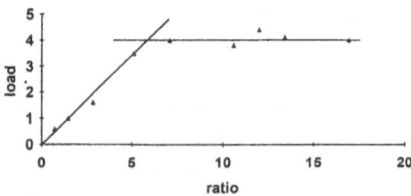

Figure 5: Number of Bengal Rose molecules encapsulated in one dendritic box as determined by UV-vis spectroscopy versus the molar ratio of Bengal Rose and dendrimer during the encapsulation reaction.

Eriochrome black T is a pH-dependent dye that is well soluble in polar solvents and can be encapsulated in the box. Due to the many (62) tertiary amines present in the interior of the box, eriochrome black T shifts its absorption spectrum from λ_{max} = 280 nm for free dye in CH_2Cl_2 to λ_{max} = 360 and 570 nm for dye in the box and in CH_2Cl_2. As soon as the absorption spectrum of free dye and encapsulated dye are different it is not possible to determine the number of molecules encapsulated, accurately, in a simple way. Since eriochrome black T is well soluble in water or acetonitrile, while the dye@box is insoluble in these solvents, we used this system to study the diffusion from the dye out of the box. Even after prolonged heating, dialysis or sonofication the water of the dispersion did not become colored due to diffusion, therefore, it was concluded that the diffusion of dye out of the box is unmeasurably slow.

By comparing the encapsulation results of a large variety of dye molecules, it became apparent that many coplanar dye molecules with an ionic group can be encapsulated into the dendritic box. For concentrated solutions of large dyes in the encapsulation reaction the maximum number of dye molecules entrapped is 4, which is related to the architecture of the dendritic box. Large three-dimensional dyes or coplanar dyes without ionic or polar groups are hard to encapsulate and only small numbers of the ratio guest per host are observed, typically around 0.1 as the average number of guests per box. Smaller polar guests can be encapsulated with maximum numbers beyond 4, but almost in all cases an integer number of 6 or 10 is found.

These results suggest that the procedure employed here produces a unimolecular compartmented structure in which guest molecules can be encapsulated and for which the diffusion out of the box is unmeasurably slow.

OPTICAL ACTIVITY OF THE DENDRITIC BOX

Since the modification reaction of the poly(propylene imine) dendrimers is performed with enantiomerically pure amino acids, it is of interest to study the chiroptical properties of the box and the guest@box systems [13,14]. Much to our surprise, we noticed that the optical activity of the DAB-*dendr*-(NH-t-BOC-L-Phe)n decreases drastically on going from dendrimers of the first generation with four end groups ($[\alpha]_D$ = -11; c=1, $CHCl_3$) to the dendritic box of the fifth generation ($[\alpha]_D$ = -0.1; c=1, $CHCl_3$) with 64 end groups. This decrease in optical activity is not due to (partial) racemisation of the amino acids employed, as was demonstrated with HPLC using a chiral stationary phase. The specific optical rotations as a function of generation are given in figure 6. A more thorough investigation employing a variety of different amino acid derivatives revealed that this decrease of optical rotation with increasing generation is a general phenomenon for all of the t-BOC-protected amino acids used [13]. Circular dichroism and optical rotatory dispersion measurements confirmed the results of the specific optical rotations. Using model sytems we investigated the solvent dependence of the lower generations and found that for the L-Phe derivative the optical rotation is strongly influenced by the solvent. The optical rotations varied from $[\alpha]_D$ = 7.3 (c=1, toluene) to $[\alpha]_D$ = -6.4 (c=1, acetonitrile).

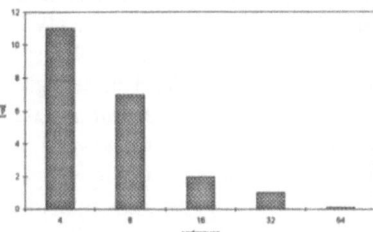

Figure 6: Specific optical rotations at the Sodium D-line for the t-BOC-L-Phe modified poly(propylene imine) dendrimers as measured in $CHCl_3$.

In order to investigate the importance of both the amide and the carbamate functionality in the end groups, a system is synthesized in which the carbamate group is replaced by an acetal moiety, while the shape is almost constant (figure 7). When the model compound of the propylamine and the endgroup was exerted to optical rotation measurements in various solvents, it was shown that this model compound showed only a marginal solvent dependency as the $[\alpha]_D$ varied at values between 40 and 60. The specific optical rotations of the various generations of acetal-modified poly(propylene imine) dendrimers were measured and proved to be independent on the generation [15].

Figure 7: The dendritic box versus a modification in which the carbamate group is replaced by an acetal moiety.

In order to explain the peculiar optical behavior of the dendritic box, with its solid-like shell, it is first necessary to explain the difference in solvent dependence of the model compounds on the optical activity. A strong solvent, concentration-, or temperature-dependence on the optical rotation of organic compounds has been observed already about a century ago [16], but a detailed rational explanation is still lacking [17]. It is assumed that different solvents give rise to different distributions of conformations, which sometimes (but not necessarily) leads to large differences in optical rotations. If in the dendritic shell of the box several conformations are frozen-in, this will lead to a kind of average optical activity. In the case of the dendritic box this will tend to a vanishing optical rotation and for the acetal-modified dendrimers to a nearly constant value of about 42. Hence, the highly-dense packing of end groups in the multiple-hydrogen bonded shell of the dendritic box gives rise to different frozen-in conformations, leading to an internal compensation of optical activity.

Stimulated by the observation of induced chirality of dyes dissolved into chiral bilayers and micelles [18], the circular dichroism (CD) spectra of a variety of dyes encapsulated in the dendritic box have been recorded. Induced circular dichroism spectroscopy is based on the transfer of chirality from the environment to an achiral dye and could therefore be applicable to these boxes. The vanishing optical activity is caused by a compensation effect and local optical activity is still thought to be present. In figure 8 the results are given for two samples of Bengal Rose@DAB-dendr-(NH-t-BOC-L-Phe)$_{64}$ with on the average 1 and with 4 molecules of Bengal Rose per dendritic box. Although both samples show identical UV spectra, a dramatic difference is observed in the induced CD spectra of both samples. The dendritic box with 1 molecule of Bengal Rose encapsulated exhibits an induced CD spectrum related to the UV spectrum, in which all bands possess a negative Cotton effect. However, an exciton-coupled spectrum is observed when on average 4 molecules of Bengal Rose are encapsulated in a single dendritic box. This exciton coupling indicates the close proximity of chromophores with a certain fixed orientation [19]. All explanations for the induced CD observed are speculative, however, it is reasonable to assume that some kind of chirality is present in the cavities of the dendritic box, despite the vanishing optical activity of the shell.

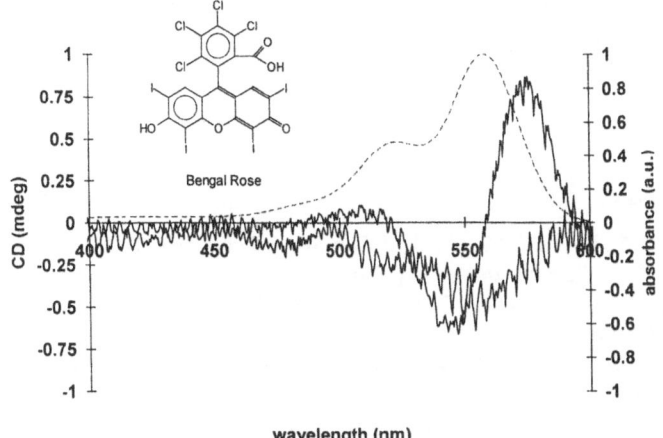

Figure 8: UV (A) and CD spectra of Bengal Rose@DAB-*dendr*-(NH-t-BOC-L-Phe)₆₄ containing 1 (C) and 4 (B) molecules of Bengal Rose.

SHAPE-SELECTIVE LIBERATION OF ENCAPSULATED GUESTS

Sofar the dendritic box is used to encapsulate guest molecules into the internal cavities present. The rigid, densely-packed shell of the DAB-*dendr*-(NH-t-BOC-L-Phe)₆₄ limits the diffusion out of the box of almost all guest molecules studied sofar. Obviously, it is difficult to determine the diffusion of solvent molecules accurately, but all experimental data available sofar show that small molecules like CH_2Cl_2 can penetrate through the rigid shell. If a dendritic box is made from the t-BOC protected glycine amino acid a semi-permeable box is made [20]. This idea of tuning the density of the shell by decreasing the size of the end groups has been used to obtain a shape-selective liberation of guests form the dendritic box made from L-Phe [21].

After encapsulation of four molecules of Bengal Rose and 8-10 molecules of *para*-nitrobenzoic acid together in a dendritic box, hydrolysis of the t-BOC groups with formic acid (95% HCOOH, 16h) was performed. Subsequent dialysis (5% water in aceton) of the reaction mixture yielded a perforated dendritic box in which only the four molecules of Bengal Rose are ent rapped, whereas all *para*-nitrobenzoic acid was dissolved in the aceton/water mixture. Bengal Rose cannot be liberated from the perforated box, not even after the addition of 12 N hydrochloric acid. However, hydrolysis of the outer shell using 12 N HCl under reflux for 2h liberated Bengal Rose after dialysis (100% water) and the starting poly(propylene imine) dendrimer is recovered in 50-70% yield. By applying this two-step hydrolysis procedure to a variety of different mixtures of guest molecules it was shown that this shape-selective liberation is a general principle [20]. Furthermore by changing the amino acids in the shell and the protecting group of the amino acid it proved to be possible to fine-tune this pathway of liberation completely.

CONCLUSIONS

We have discussed the synthesis of dendritic boxes possessing a unimolecular compartmented structure in which guest molecules are physically locked. Evidence is presented that the encapsulation is dominated by the architecture of the dendrimer and that some kind of supramolecular ordering is present. Furthermore, a shape-selective liberation of guests can be accomplished by a two-step process. Further research is in progress to investigate the scope and limitations of this new guest-host system. It is envisaged that with this dendritic box, we can enter the field of modular chemistry.

Acknowledgement: The authors like to thank many of their collegues at the Eindhoven University and DSM Research for stimulating discussions, experimental assistance and enthousiasm to enter the field of dendrimers. Many of the names are found in the list of references to the original literature.

References and notes:

1. Tomalia D.A., Naylor A.M., Goddard W.A. III, Angew Chem, **102**, 119 (1990).
2. Fréchet J.M.J., Science, **263**, 1710 (1994).
3. NewKome G.R., Moorefield C.N., Baker G.R., Saunders M.J., Grossman S.H., Angew. Chem., **103**, 1207 (1991).
4. Tomalia D.A., Baker H., Dewald J., Hall M., Kallos G., Martin S., Roeck J., Ryder J., Smith P., Polymer Journal, **17**, 117 (1985).
5. Wooley K.L., Hawker G.J., Fréchet J.M.J., J. Am. Chem. Soc., **115**, 11496 (1993).
6. Tomalia D.A., Hurst D.A., In: Weber E. (ed), Topics in Current Chemistry, Springer-Verlag, Berlin, **165**, 93 (1993).
7. Buhleier E., Wehner W., Vögtle F., Synthesis, 155 (1978).
8. Maciejewski M., J. Macromol. Sci. - Chem., **A17**, 689 (1982).
9. Jansen J.F.G.A., de Brabander - van den Berg E.M.M., Meijer E.W., Science, **266**, 1226 (1994).
10. De Brabander - van den Berg E.M.M., Meijer E.W., Angew. Chem., **105**, 1370 (1993).
11. Jansen J.F.G.A., Meijer E.W., manuscript in preparation.
12. Jansen J.F.G.A., Janssen R.A.J., de Brabander - van den Berg E.M.M., Meijer E.W., Adv. Mat., **7**, 561 (1995).
13. Jansen J.F.G.A., Peerlings H.W.I., de Brabander - van den Berg E.M.M., Meijer E.W., Angew. Chemie. Int. Ed. Eng., **34**, 1206 (1995).
14. Jansen J.F.G.A., de Brabander - van den Berg E.M.M., Meijer E.W., Recl. Trav. Chim. Pays-Bas, **114**, 225 (1995).
15. Peerlings H.W.I., Jansen J.F.G.A., de Brabander - van den Berg E.M.M., Meijer E.W., PMSE, **73**, 342, ACS meeting Chicago (1995).
16. Winther C., Z. Phys. Chem., **60**, 621 (1897).
17. Eliel E.L., Wilen S.H., In: Stereochemistry of Organic Compounds, J Wiley & Sons Inc., New York, 1076 (1994).
18. Kurzitake T., Nakashima N., Moritsu K., Chem. Lett., 1347 (1980).
19. Bosman A.W., Janssen R.A.J., Meijer E.W., unpublished results.
20. Jansen J.F.G.A., de Brabander - van den Berg E.M.M., Meijer E.W., J. Am. Chem. Soc., **117**, 4417 (1995).

New Macromolecular Architectures and Functions Through Macromolecular Recognition by Cyclodextrins

A. Harada and M. Kamachi

Department of Macromolecular Science, Faculty of Science,
Osaka University, Toyonaka, Osaka, 560 Japan

Abstract: Cyclodextrins(CDs) were found to form inclusion complexes not only with poly(ethylene glycol) (PEG) but also poly(propylene glycol) (PPG), poly(methyl vinyl ether) (PMVE), and poly(oxytrimethylene) (POT) to give crystalline compounds in high yields. The complex formation is highly selective. α-CD formed complexes with PEG and POT, although it did not form complexes with PPG and PMVE. γ-CD formed complexes with PMVE, although α- and β-CD did not form complexes with PMVE. The complexes are stoichiometric compounds and characterized by IR, ^{1}H NMR, ^{13}C NMR, ^{13}C CP/MAS NMR, ^{13}C PST/MAS NMR spectra and X-ray, thermal, and elemental analyses. CDs were also found to form inclusion complexes with hydrophobic polymers, such as oligoethylene and polyisobutylene. α-CD formed complexes with oligoethylene of molecular weight less than 1000. β-CD and γ-CD formed complexes with polyisobutylene (PIB), although α-CD did not form complexes with PIB of any molecular weight. The yield of the complex with β-CD decreased with increasing the molecular weight of PIB. In contrast, the yield of the complex with γ-CD with PIB increased with increasing the molecular weight. The chain length selectivities are reversed.

INTRODUCTION

In recent years, much attention has been focused on the design and construction of nano-scale structures [1]. Polymers with unique structures, such as double helical polymers [2], dendrimers [3], and bilayer structures [4], have been prepared and attracted much attention because of their unique properties and functions. We have been pursuing a new method to create new supramolecular assemblies from polymers and interacting smaller molecules.

We have chosen cyclodextrins as molecular parts for constructing supramolecular assemblies with polymers. Cyclodextrins are cyclic

M. Kamachi · A. Nakamura (Eds)
New Macromolecular Architecture and Functions
Proceedings of the OUMS '95 Toyonaka, Osaka, Japan, 2-5 June, 1995
© Springer-Verlag Berlin Heidelberg 1996

molecules consisting of 6 to 8 glucose units linking through α-1-4, glycosidic linkages. They are called α-, β-, and γ-cyclodextrin (CD), respectively. They are known to form inclusion complexes with a wide variety of low molecular weight compounds, ranging from nonpolar hydrocarbons to polar carboxylic acids and amines. Since cyclodextrins were discovered, a great number of reports (more than 10 000 papers) have been published on cyclodextrins. However, studies on the inclusion properties of cyclodextrins were limited to those of low molecular weight compound [5-7]. There were no reports on the complex formation of cyclodextrins with polymers when we started our work in early 1980's. We have started our project on complex formation between polymers and cyclodextrins. Previously we have reported that cyclodextrins form inclusion complexes with poly(ethylene glycol) (PEG) to give stoichio-metric compounds in crystalline state in high yield [8-14]. This is the first observation that cyclodextrins form complexes with polymers. We have prepared polyrotaxanes from these complexes [15-18].

We have found that CDs formed complexes not only with hydrophilic polymers but also hydrophobic polymers and oligomers, such as oligoethylene [19] and polyisobutylene [20]. We describe here the complex formation between cyclodextrins and some water soluble polymers and hydrophobic polymers.

COMPLEX FORMATION BETWEEN CYCLODEXTRINS AND HYDROPHILIC POLYMERS

Table 1 shows the results of the formation of the complexes between hydrophilic polymers with the same composition of -(C3H6O)n- and cyclodextrins. α-CD (diameter of the cavity : 4.5 Å) formed complexes with POT [21], which has the smallest cross-sectional area in high yield, although γ-CD did not form complexes with POT of any molecular weight. β-CD (diameter of the cavity : 7.0 Å) and γ-CD (diameter of the cavity : 8.5 Å) formed complexes with PPG [22,23], which has larger cross-sectional area than POX, although α-CD did not form complexes with PPG of any molecular weight. It should be noted that γ-CD specifically formed complexes with PMVE [24], which has the same composition as those of PPG and POT but carries a methoxy group as a side chain, although α- and β-CD did not form complexes with PMVE of any molecular weight.

Table 1

Complex Formation of CDs with -(C3H6O)n- Polymers

Polymer	Structure	MW	Yield(%)		
			α-CD	β-CD	γ-CD
POx	-(CH2CH2CH2O)-	1000	94	47	0
PPG	-(CH2CHO)- CH3	1000	0	96	80
PMVE	-(CH2CH)- OCH3	20000	0	0	80

These results indicate that the relative sizes of cyclodextrins and the cross-sectional area of the polymers are important in the complex formation.

EFFECTS OF MOLECULAR WEIGHT OF POLYMERS ON THE COMPLEX FORMATION

Figure 1 shows results of the complex formation of β-CD with PPG of various molecular weights. β-CD did not form complexes with the low molecular weight analogs such as propylene glycol, dipropylene glycol, and tripropylene glycol. β-CD formed complexes with PPG with molecular weight higher than 400. Yields of the complexes increase with increasing molecular weight. The complexes were obtained almost quantitatively between β-CD and PPG with molecular weight 1000. Then the yields decrease with increase in molecular weight. Although this result is consistent with that of the complex formation between α-CD and POT, it is different from the complex formation between α-CD and PEG, where the yields increase with increasing molecular weight and reach saturation. This may be due to the fact that PPG is more hydrophobic than PEG owing to the methyl group of the main chain. γ-CD also formed complexes with PPG, even with PPG of low molecular weight in good yield. α-CD did not form complexes with PPG of any molecular weight, although it formed complexes with PEG of various molecular weight to give crystalline complexes in high yields. It is in contrast, β-CD did not form complexes with PEG of any molecular weight. An α-CD cavity is too small for PPG to penetrate due to steric hindrance by methyl groups on the main chain.

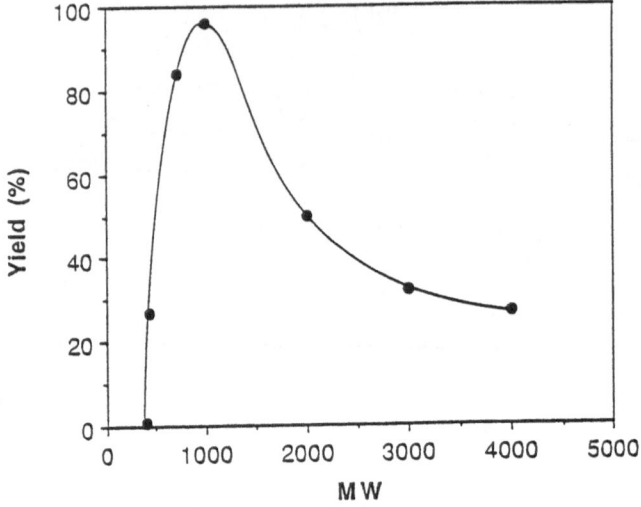

Fig. 1. Yields of complexes of β-CD with PPG as a function of the MW of PPG

Figure 1 shows that a minimum chain length is required for the formation of crystalline complexes with β-CD. The same behavior was observed in the formation of crystalline complexes of PEG with α-CD. This is thought to be characteristic of crystalline complex formation between hydrophilic polymers and cyclodextrins. This result shows the importance of cooperative effects in the complex formation. The cooperation is thought to result from the fact that a single polymer chain has many binding sites which are included by cyclodextrin molecules.

The neighboring CDs bound on a polymer chain interact with each other by forming hydrogen bonds. This view is consistent with the fact that PPG does not form crystalline complexes with 2,6-di-*O*-methyl-β-CD, 2,3,6-tri-*O*-methyl-β-CD, and water-soluble β-CD polymer. These compounds are thought to be unable to include a PPG chain to form crystalline complexes, because they cannot form hydrogen bonds due to the lack of hydroxyl groups.

Stoichiometries of the Complexes

Continuous variation plots for the complex formation and ^{1}H NMR spectra of the complexes of polymers with CDs show that the complexes are stoichiometric. PPG-β-CD complex and PPG-γ-CD complex are 2:1 (two monomer units and one CD). Stoichiometry of the POT-α-CD complex is 1.6 : 1 (CD : monomer unit) when the molecular weight of POT is less than 2000. This result indicates that a single α-CD molecule binds six atoms in the main chain and is consistent with the result on the complex formation between PEG and α-CD.

Properties

The complexes of β-CD with PPG of low molecular weight (MW=400-700) are soluble in a large amount of water, but those with PPG of higher molecular weights are sparingly soluble in water. These results are in contrast to the complexes of PEG with α-CD, which are soluble in a large amount of water or by heating. This is owing to the fact that PPG and POT are more hydrophobic than PEG. The complexes between POT and α-CD are sparing soluble in water but they are solubilized on heating or addition of urea, indicating that the complex formation is reversible phenomenon and that the hydrogen bonding plays an important role in stabilizing the complexes. The complexes are soluble in dimethylsulfoxide and dimethylformamide. The X-ray diffraction studies show that all the complexes are crystalline, in spite of the fact that PPG and POT are liquid.

BINDING MODES

Figure 2 shows the X-ray pattern of β-CD (a), the complex of β-CD with *p*-nitroacetanilide (b) and with PPG (MW=1000) (c). Saenger et al. reported that the structures of the inclusion complexes of CDs with low molecular weight compounds can be classified as "cage type" or "channel type". [7] The patterns show that all the complexes are crystalline and the pattern of the PPG complex is different from that of free β-CD and those of the complexes with small molecules, such as propanol, but similar to that of the complex with *p*-nitroacetanilide, which has been proved to have a column structure by the X-ray study of a single crystal of the complex. [25] These results indicate that in the PPG complex with β-CD, the CD shows a packing different from that in free β-CD and has channel structures as in the complex with *p*-nitroacetanilide.

112

X-ray powder pattern of the complex between α-CD and POT is the same as that of the complex between PEG and α-CD and those between valeric acid or octanol and α-CD, which have been reported to have channel type structure. Therefore, we conclude that α-CD formed a complex with POT similar to those of PEG with α-CD.

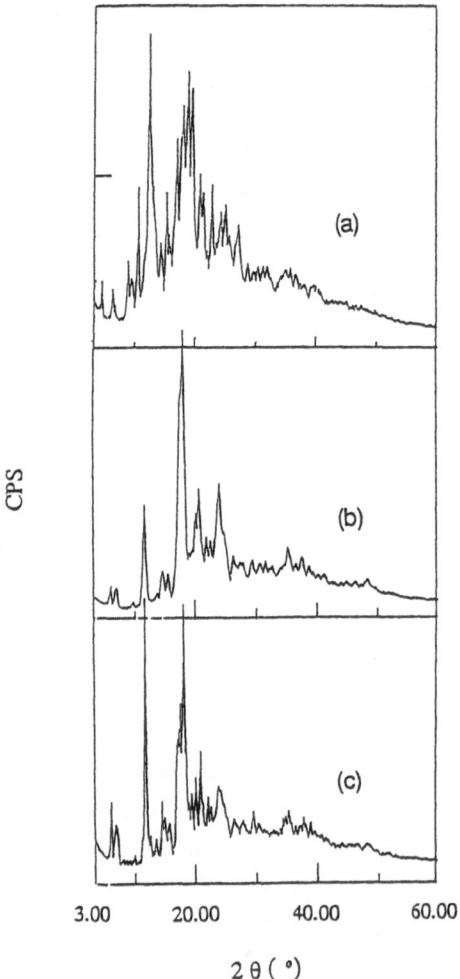

Fig. 2. X-ray diffraction patterns of β-CD (a), β-CD-*p*-nitroacetanilide complex (b), and β-CD-PPG complex (MW=1000) (c)

Figure 3 shows the CP/MAS NMR and PST/MAS NMR spectra of the complexes of β-CD and PPG (MW=1000) . β-CD assumes a less symmetrical conformation in the crystal when it does not include a guest in the cavity. In this case, the spectrum shows resolved C-1 and C-4 resonances from each of the α-1,4-linked glucose residues. Especially, two peaks at 94.3 and 96.8 ppm, which are assigned to the conformationally strained glycosidic linkage, are observed.
On the other hand, in the spectrum of the β-CD-PPG complex, the peaks at 94.3 and 96.8 ppm disappeared. Each carbon of glucose can be observed in a single peak. These results indicate that β-CD adopts a symmetrical conformation and each glucose unit of β-CD is in a similar environment. The X-ray studies of single crystals showed that β-CD adopts a symmetrical conformation when it included guests in the cavities.

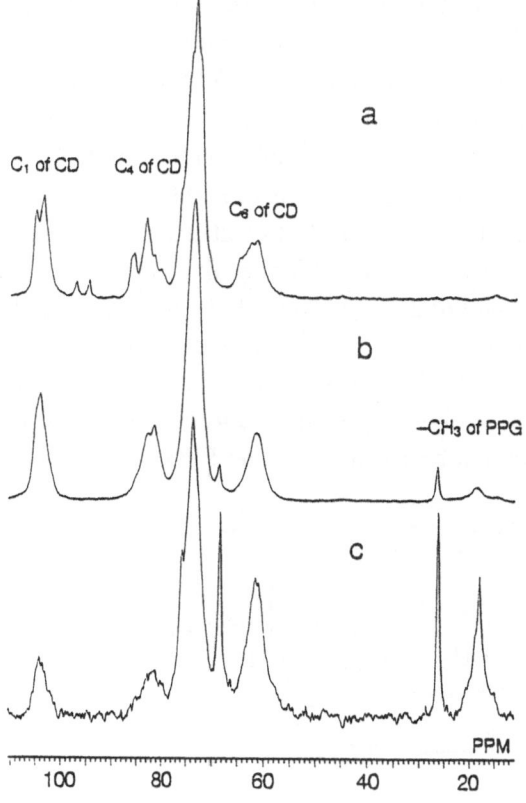

Fig. 3. ^{13}C-CP/MAS NMR spectra of β-CD (a) and the β-CD-PPG complex (b) and ^{13}C-PST/MAS NMR spectrum of β-CD-PPG complex

114

The results of CP/MAS NMR spectra of complexes and uncomplexed cyclodextrins are consistent with the results of X-ray studies. Therefore, a PPG chain is thought to be included in the cavities of cyclodextrins. Figure 3 (c) shows the ^{13}C PST/MAS NMR spectrum of β-CD-PPG complex, which gives stronger signals of relatively flexible carbons of the sample than ^{13}C CP/MAS NMR. The relative intensities of the peaks of PPG to those of β-CD in Figure 3 (c) are much stronger than those in Figure 3 (b), showing that the PPG chain is not as rigid as β-CD in the complex. These results are consistent with the view that β-CD molecules form a channel, forming the crystal frame of the complexes. PPG chain is included in the channel.

Molecular model studies show that POT chains are able to penetrate α-CD cavities and PPG chains are able to penetrate β-CD cavities, while the PPG chain cannot penetrate α-CD cavities owing to the hindrance of methyl groups on the main chain. Model studies further indicate that the single cavity of α-CD accommodates 1.5-1.6 POT unit and a β-CD cavity accommodates two propylene glycol units. The inclusion complex formation of polymers with cyclodextrins is entropically unfavorable. However, formation of the complexes is thought to be promoted by hydrogen bond formation between cyclodextrins. Therefore, the head-to-head and tail-to-tail arrangement, which results in a more effective formation of hydrogen bonds between cyclodextrins, is thought to be the most probable structure. Such structure was proved by X-ray studies on a single crystal of the complex between β-CD and *p*-nitroacetanilide and that of α-CD and tetraethylene glycol. **Figure 4** shows a proposed structure of the complex between β-CD and PPG.

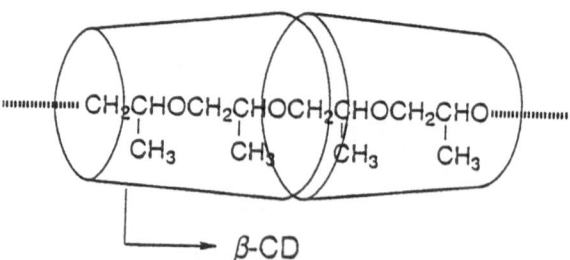

Fig. 4. Proposed structure of the complex between β-CD and PPG

COMPLEX FORMATION OF CYCLODEXTRINS WHIT HYDROPHOBIC POLYMERS

We found that cyclodextrins form complexes with not only hydrophilic polymers but also hydrophobic polymers, such as oligoethylene and polyisobutylene. α−CD forms complexes with oligoethylenes, although β- and γ-CD did not form complexes oligoethylenes under the same conditions. On the other hand, β- and γ-CD formed complexes with polyisobutylene (PIB), although α-CD did not form complexes with PIB.

Complex Formation of Oligoethylene with α-Cyclodextrin [19]

Oligoethylene (OE) was found to form inclusion complexes with α-CD not only from aqueous solutions of α-CD but also from DMF solutions of α-CD to give stoichiometric compounds in a crystalline state. The stoichiometries of the complexes are 3:1 (ethylene unit:α-CD). The X-ray diffraction studies and ^{13}C CP/MAS and PST/MAS NMR spectra suggest that the OE chain is included in α-CDs and the OE backbone in the complexes is more flexible than that in uncomplexed state.

Complex Formation of Polyisobutylene with Cyclodextrins [20]

α-CD did not form complexes with PIB of any molecular weight. β-CD and γ-CD formed complexes with PIB. The yields of the complexes with β-CD decreased with increase in the molecular weight of PIB. In contrast, the yields of the complexes with γ-CD increased with increase in the molecular weight and the complexes were obtained almost quantitatively with PIB of molecular weight of 1000. The chain length selectivity is totally reversed between β-CD and γ-CD.

CONCLUSION

Cyclodextrin were found to form inclusion complexes not only with low molecular weight compounds but also with hydrophilic and hydrophobic polymers to give stoichiometric compounds. The selectivities shown by cyclodextrins toward polymers are much higher than for low molecular weight compounds. This is due to the fact that the guest polymers have a lot of recognition sites. This is one of the reasons why the living systems are composed of many kinds of macromolecules. This kind of complex formation can be utilized to create new supramolecular architectures and functions [26,27].

116

References

[1] Whitesides GM, Mathias JP, Seto CT (1991) Science 254: 1312

[2] Bell TW, Jousselin H (1994) Nature 367: 441

[3] Mekelburger HB, Jaworek W, Vogtle F (1992) Angew. Chem. Int. Ed. Engl. 31: 1571

[4] Asakuma S, Okada H, Kunitake T (1991) J. Am. Chem. Soc. 113: 1749

[5] Bender ML, Komiyama M (1978) cyclodextrin Chemistry, Springer-Verlag, Berlin

[6] Szejtli J Cyclodextrins and Their Inclusion Complexes, Akademiai Kiado, Budapest

[7] Saenger W (1976)Jerusalem Symp. Quantum Chem. Biochem. Ed. Pullman EB, Reidel Co., Dordrecht

[8] Harada A (1993) Polym News 18:358

[9] Harada A, Kamachi M (1990) Macromolecules 23: 2821

[10] Harada A, Li J, Kamachi M (1993) Proc. Jpn. Acad. 69, SerB:39

[11] Harada A, Li J, Kamachi M (1993) Macromolecules 26: 5698

[12] Harada A, Li J, Kamachi M (1994) Macromolecules 27: 4538

[13] Li J, Harada A. Kamachi M (1994) Polym. J. 26:1019

[14] Harada A, Li J, Kamachi M (1994) Ordering in Macromolecular Systems, 69

[15] Harada A, Li J, Kamachi M (1992) Nature 356: 325

[16] Harada A, Li J, Kamachi M (1993) Carbohydr. and Carbohydr. polym. 25: 266

[17] Harada A, Li J, Nakamitsu T, Kamachi M (1993) J. Org. Chem. 58: 7524

[18] Harada A, Li J, Kamachi M (1994) J. Am. Chem. Soc. 116: 3192

[19] Li J, Harada A. Kamachi M (1994) Bull. Chem. Soc., Jpn., 67: 2808

[20] Harada A, Li J, Suzuki S, Kamachi M (1993) Macromolecules 26: 5267

[21] Harada A, Okada Y, Kamachi M Acta Polym., *in press*

[22] Harada A, Kamachi M (1990) J. Chem. Soc., Chem. Commun., 1322

[23] Harada A, Okada Y, Li J, Kamachi M Macromolecules *in press*

[24] Harada A, Li J, Kamachi M (1993) Chem. Lett., 237

[25] Harding MM, MacLennan JM, Paton RM (1978) Nature 274: 621

[26] Harada A, Li J, Kamachi M (1993) Nature 364: 516

[27] Harada A, Li J, Kamachi M (1994) Nature 356: 325

Novel Unimolecular Micelles of Hydrophobically Modified Polyelectrolytes: Synthesis, Characterization, and Functions

Y. Morishima and M. Kamachi

Department of Macromolecular Science, Faculty of Science,
Osaka University, Toyonaka, Osaka 560, Japan

Abstract: Random copolymers of sodium 2-acrylamido-2-methylpropanesulfonate (AMPS) and methacrylamides N-substituted with bulky hydrophobic groups with cyclic structures such as cyclododecyl (Cd), adamantyl (Ad), and 1-naphthyl (1-Np) groups undergo intrapolymer self-organization in aqueous solution and form unimolecular micelles (unimers) independent of the polymer concentrations. These unimers are highly compact as indicated by light scattering and NMR relaxation times. Hydrophobic chromophores can be tightly encapsulated in the hydrophobic cluster in the unimer by covalently incorporating into the hydrophobically modified polysulfonates. The chromophores compartmentalized in the unimers are completely isolated from one another in highly constraining nonpolar microenvironments and are protected from the aqueous phase, leading to a large modification of the photophysical and photochemical behavior.

1. INTRODUCTION

Water-soluble amphiphilic polymers have been the subject of extensive studies over the past decade because of both scientific and industrial interests [1,2]. In hydrophobically modified polyelectrolytes, hydrophobic interaction competes with electrostatic repulsion, and therefore the balance of the contents of the hydrophobic and charged units in a polymer chain is a critical factor to determine whether or not the polymer collapses owing to cooperative hydrophobic association. In general, intrapolymer hydrophobic association may be dominant in dilute polymer solutions, whereas interpolymer association may preferably occur in a semi-dilute or concentrated regime. In the case of polymers of a strong tendency for intrapolymer association, unimolecular micelles (unimers) may be formed independent of the polymer concentration. In contrast, polymers of a strong propensity for interpolymer association leads to multipolymer aggregates that may or may not accompany phase separation.

This review deals with the self-organization of polysulfonates modified with bulky hydrophobes, characterization of their unimers, and some photochemical behavior of the unimers functionalized with chromophores.

2. SELF-ORGANIZATION OF HYDROPHOBICALLY MODIFIED POLYELECTROLYTES

If the content of a hydrophobe in a hydrophobically modified polyelectrolyte is above a certain critical content, the polymer would undergo self-organization in aqueous solution. The first step of the self-organization may be intrapolymer hydrophobic association to form micelle units along a polymer chain (Figure 1). However, the conformation thus formed, which may be viewed as a

M. Kamachi · A. Nakamura (Eds)
New Macromolecular Architecture and Functions
Proceedings of the OUMS '95 Toyonaka, Osaka, Japan, 2-5 June, 1995
© Springer-Verlag Berlin Heidelberg 1996

"second-order structure", may not be stable because a significant portion of the surface of the hydrophobic cluster is exposed to water. Consequently, as schematically illustrated in Figure 1, the micelle units may congregate to form a higher-order structure and thus the polymer may end up with a micelle made up with a single macromolecule (unimer).

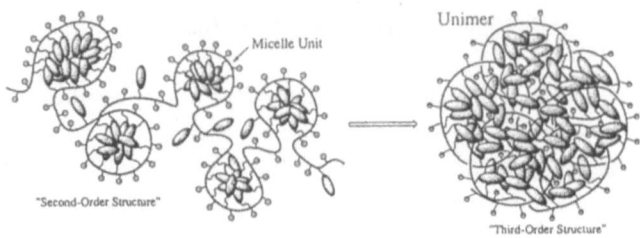

Figure 1. A schematic model of unimers.

In general, the hydrophobic association may not necessarily occur only in an intrapolymer mode. If interpolymer hydrophobic association occurs, then multipolymer aggregates would be formed instead of the unimers. However, whether the intrapolymer association predominates over the interpolymer association depends primarily on the chemical structure (the first-order structure) of the hydrophobically modified polyelectrolyte. Recently, McCormick *et al.* [3] and Morishima *et al.* [4] have independently shown that the sequence distribution of charged units and hydrophobic units along the polymer chain is an important factor to determine the mode of hydrophobic association; block sequences have a strong tendency for interpolymer association [3], whereas random and alternating sequences tend to associate in an intrapolymer mode. Another important structural factors for unimer-forming polymers is that the hydrophobes should be bulky with cyclic structures such as cyclododecane, adamantane, and naphthalene, and their contents should be higher than ca. 30 mol % [4]. Furthermore, amide spacer bonds may be important because they form hydrogen bond networks that may contribute to retaining the compact unimer structure [4].

These unimers are very different from the classical surfactant micelles in that (1) all charged and hydrophobic groups are covalently linked to the polymer backbone, (2) the unimers are "static" in nature as oppose to the "dynamic" nature of the surfactant micelles which exist in equilibrium between association and dissociation, and (3) the unimer structure is retained as such even at very low and high concentrations.

3. SYNTHESIS OF UNIMER-FORMING AMPHIPHILIC POLYELECTROLYTES

Unimer-forming amphiphilic polyelectrolytes can be synthesized by free-radical copolymerization of 2-acrylamido-2-methylpropanesulfonate (AMPS) and methacrylamides substituted with bulky hydrophobes of cyclic structures such as *N*-cyclododecylmethacrylamide (CdMAm), *N*-(1-adamantyl)methacrylamide (AdMAm), and *N*-(1-naphthylmethyl)methacrylamide (1NpMAm) [5]. The composition of the unimer-forming copolymers ranges 40-70 mol % in the AMPS unit and 30-60 mol % in the hydrophobic monomer unit.

For spectroscopic characterization, these polymers can be labeled with a small amount of pyrenyl

(Py) moieties by terpolymerization using *N*-(1-pyrenylmethyl)methacrylamide (PyMAm) [5].

These polymerizations are characterized as the "ideal copolymerization" where the monomer reactivity ratios are practically unity, resulting in copolymer compositions equal to monomer feed compositions and the completely random distribution of the monomer units along the polymer chain [5,6]. Therefore, one can determine the terpolymer compositions directly by the molar ratios of the monomers in feed, which allows one to prepare co- and terpolymers with well defined compositions and completely random distributions of the monomeric units along the polymer chain. The terpolymers with up to ca. 60 mol% hydrophobic monomer units are completely soluble in water to give optically clear solutions.

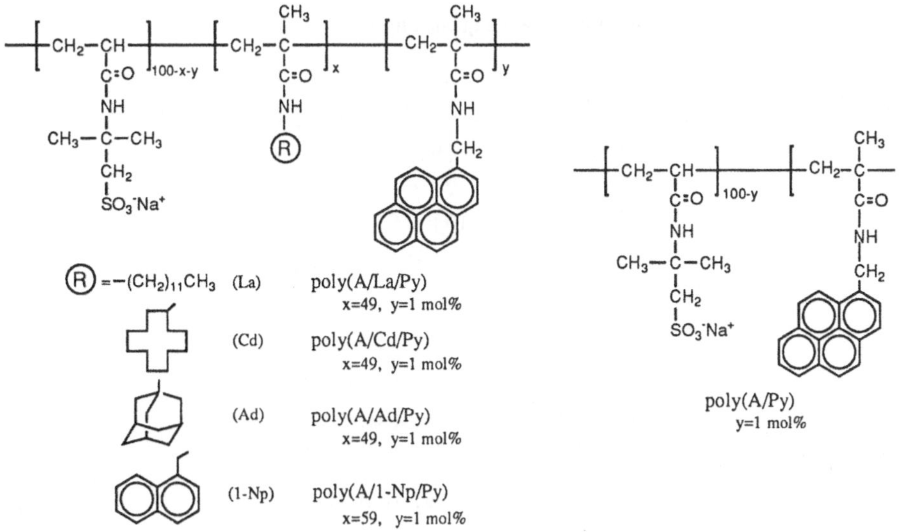

Scheme 1

In the text, the terpolymers are denoted as poly(A/R/P), where A, R, and P represent the AMPS, hydrophobe, and label units, respectively (Scheme 1).

4. CHARACTERIZATION OF UNIMERS

Spectroscopic methods, such as fluorescence, NMR, and IR spectroscopies, provide microscopic information on the structure and dynamic behavior in a short dimensional range, while scattering methods, such as static light scattering (SLS), dynamic light scattering (DLS), and small-angle X-ray scattering (SAXS) techniques, provide information based on a medium or long range structure. Therefore, use of a variety of techniques in combination is of importance for characterization.

It is known that the ratio of the third to the first vibronic bands (I_3/I_1) in pyrene fluorescence spectra depends on the polarity in microenvironments where pyrene exists [7], I_3/I_1 values being larger in less polar media. The I_3/I_1 ratios of Py-labeled terpolymers poly(A/La/Py), poly(A/Cd/Py), and poly(A/Ad/Py) (Scheme 1) are 0.80, 0.83, and 0.76, respectively, all larger than the value of

0.59 for the reference copolymer poly(A/Py). This indicates that the Py labels are buried in the hydrophobic clusters in the terpolymers [5].

The encapsulation of the Py labels in the hydrophobic clusters and their protection from the aqueous phase are also indicated by a sharp suppression of fluorescence quenching by Tl^+ ions [4]. The fluorescence of pyrene is known to be quenched by Tl^+ due to an external heavy atom effect that requires a short range interaction [8]. In the reference copolymer poly(A/Py) which is in an open chain conformation, Tl^+ ions are electrostatically concentrated in the vicinity of the anionic polymer chain and can come into contact with the Py labels. Thus, the Py fluorescence in the reference copolymer is efficiently quenched by Tl^+. In contrast, because the Py labels in the terpolymers are buried inside the hydrophobic clusters, Tl^+ ions cannot reach the Py sites, giving rise to a sharp suppression of the fluorescence quenching.

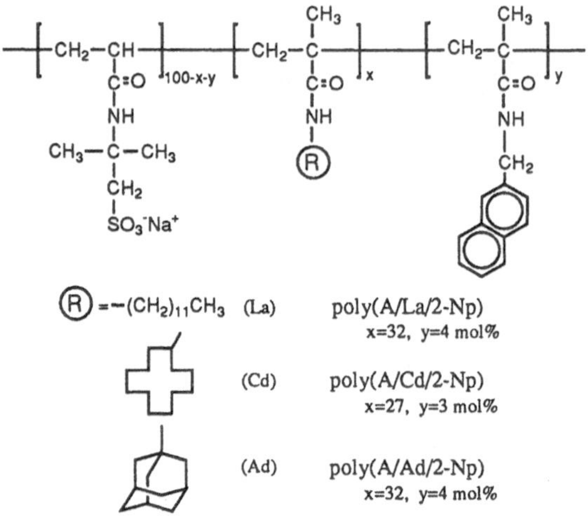

Scheme 2

Interpolymer nonradiative energy transfer (NRET) technique has been successfully applied to the La-, Cd-, and Ad-containing polysulfonates to examine whether the hydrophobic association is an intra- or interpolymer event [4]. For this experiment the La-, Cd-, and Ad-containing polymers labeled with the 2-naphthyl (2-Np) groups (Scheme 2) have been employed along with the corresponding Py-labeled polymers (Scheme 1). Aqueous solutions of the mixture of the 2-Np labeled polymer and the Py-labeled polymer, each possessing the same aliphatic hydrophobes, were irradiated at 290 nm, and fluorescences from the 2-Np and Py labels were monitored at 340 and 395 nm, respectively as a function of the total polymer concentration [4]. The 2-Np labels can be predominantly excited at 290 nm. In the case of the La-containing polymer system, the intensity of the Py fluorescence increases significantly with increasing the total polymer concentration following onset of the Py fluorescence at ca. 0.2 wt %. In contrast, the Cd- and Ad-containing polymer systems show no such increase in the Py fluorescence until the concentration is increased to ca. 7 wt %. If the self-association of the hydrophobes is a completely intrapolymer event to form a unimer, the fluorescence labels are confined to the hydrophobic clusters of the individual polymers.

Therefore, in an aqueous solution of a mixture of the 2-Np-labeled and Py-labeled polymers possessing the same aliphatic hydrophobes, the possibility of NRET from the singlet excited 2-Np label in one unimer to the Py label in another unimer should be precluded. On the other hand, if the hydrophobic association is an interpolymer event, it should be possible for the 2-Np and Py labels to come close to each other within the Förster radius (R_0=2.86 nm for NRET from 2-methylnaphthalene to pyrene [9]), which allows NRET to occur. If the interpolymer NRET takes places, fluorescence from the Py label should be observed when the 2-Np label is selectively excited.

Therefore, the observations above indicate that the Cd and Ad residues have much stronger tendency for intrapolymer association than does the La residue. In the La-containing polymer, the intrapolymer hydrophobic association occurs only at concentrations below ca. 0.2 wt %. In contrast, in the Cd- and Ad-containing polymers, the intrapolymer association is dominant and the polymers exist as the unimers over a concentration range < ca. 7 wt %. These experiments lead to an important conclusion that the mode of the hydrophobic association is very different depending on whether the structure of the hydrophobes is cyclic (Cd or Ad) or linear (La), although the numbers of the carbon atoms in these hydrophobes are more or less the same.

The NMR relaxation techniques are a useful tool to look into local segment motions in polymers [10-12]. Spin-lattice relaxation times (T_1) and spin-spin relaxation times (T_2) were determined for poly(A/La/Py) and poly(A/Cd/Py) in D$_2$O at 25 °C [15]. The T_2 value for the Cd methylene protons in the Cd-containing terpolymer was estimated to be 4 ms, being significantly smaller than T_2=34 ms for the La methyl and T_2=15 ms for the La methylene protons in the La-containing terpolymer. This indicates that the mobility of the Cd groups in the clusters is much more restricted than that of the La groups. The T_1 value for the Cd methylene in the Cd-containing terpolymer is estimated to be 904 ms, being much longer than T_1=438 ms for the La methylene in the La-containing terpolymer. The large T_1 value, along with the small T_2 value, for the Cd methylene in poly(A/Cd/Py) is indicative of highly restricted motions of the Cd groups, as compared with the La groups in poly(A/La/Py), as a result of the cluster formation.

In the amphiphilic terpolymers and reference copolymer shown in Scheme 1, the hydrophobic pendant groups are linked to the backbone via amide spacer bonds. In IR spectra measured as KBr pellets of freeze-dried samples, these polymers show characteristic IR absorption bands due to ν(C=O) and δ(NH) of the amide bond. The La-containing terpolymer exhibits a lower-wavenumber shift by 8 cm^{-1} in the δ(NH) band as compared with that of the reference copolymer, while the Cd- and Ad-containing terpolymers show much larger shifts of ca. 30 cm^{-1}. These lower-wavenumber shifts indicate the presence of the hydrogen bonds between the amide spacer bonds in the terpolymers, and the hydrogen bonds are formed more strongly in the Cd- and Ad-containing terpolymers than are in the La-containing terpolymer. A characteristic feature of the unimer of the amphiphilic polysulfonates in aqueous solution is that the hydrodynamic volume is extremely small for their high molecular weights.

In Table I are listed static light scattering (SLS) and dynamic light scattering (DLS) data for poly(A/La/Py), poly(A/Cd/Py), poly(A/Ad/Py), and poly(A/1-Np/Py) in 0.1 M NaCl solutions [4]. Stokes radii (R_s) determined by DLS are extremely small for their molecular weights. This is an experimental manifestation of the highly compact conformation of the unimers in aqueous solution. The unimer of poly(A/Cd/Py) is particularly compact, as indicated by the highest ratio of mass to dimension ($M_W=5.1 \times 10^5$ versus $R_S=5.5$ nm). The stokes radius stays fairly constant independent of the concentration, which is important evidence for unimers.

Table I. Light scattering and SAXS data for the La-, Cd-, and Ad-containing terpolymers in aqueous solution [4]

polymer	$M_W{}^a$	$R_S{}^b$ (nm)	d^c (nm)
poly(A/La/Py)	1.2×10^5	7.0	...
poly(A/Cd/Py)	5.1×10^5	5.5	11
poly(A/Ad/Py)	3.5×10^4	6.2	7.3
poly(A/1-Np/Py)	1.3×10^5	...	...

    a. Weight average molecular weight determined by SLS.

    b. Average Stokes radius determined by DLS.

    c. Spacing calculated from the scattering angle in SAXS.

Concentrated aqueous solutions of the Cd-containing and Ad-containing terpolymers show X-ray scattering peaks in the small angle region, while dilute and semi-dilute solutions show no such scattering. In SAXS of poly(A/Cd/Py) and poly(A/Ad/Py) in a concentrated (ca. 10 wt %) aqueous solution, scattering peaks were observed at $2\theta=0.8$ and $1.2°$ for the Cd- and Ad-containing terpolymers, respectively. From these peak angles the spacings are calculated to be 11 and 7.3 nm for the Cd-containing and Ad-containing terpolymers, respectively [4]. A spacing of 11 nm for the Cd-containing terpolymer coincides with the Stokes diameter determined by DLS (Table I). These scattering peaks can be interpreted to be due to closely packed unimer particles in the concentrated aqueous solutions. The Cd- and Ad-containing terpolymers remain as the unimers even at high concentrations as discussed in the previous section, and the unimers exist independently (without interpenetrating) even at the high concentration. In the case of poly(A/La/Py), no scattering peaks were recognized in aqueous solution in the small angle region. This is because the La-containing terpolymer can only exist as the unimer at concentrations below ca. 0.2 wt % as described in the previous section, and the interpolymer association occurs at higher concentrations.

5. PHOTOCHEMICAL BEHAVIOR OF FUNCTIONALIZED UNIMERS

If a small amount of hydrophobic chromophores is covalently incorporated into the unimer of hydrophobically modified polysulfonates, the chromophores are tightly encapsulated in the hydrophobic cluster in the unimer ("compartmentalization"). The chromophores compartmentalized in the unimers are completely isolated from one another in highly constraining nonpolar microenvironments, and are protected from the aqueous phase [5]. Unlike the conventional molecular assembly systems, the unimer induces a large modification of the photophysical and photochemical behavior of the compartmentalized dyes [13-21].

Very rapid and efficient energy migration and trapping occur within the unimer consisting of naphthalene clusters and pyrene labels [16]. A terpolymer of 40 mol % AMPS , 59 mol % 1NpMAm, and 1 mol % PyMAm , poly(A/1-Np/Py) (Scheme 1), forms a unimer in water, the Py species being compartmentalized in the 1-Np clusters. From the weight average molecular weight

(Table I) and the compositions of the 1-Np and Py units in the terpolymer, it can be roughly estimated that there exist 300 1-Np and 5 Py units in a unimer.

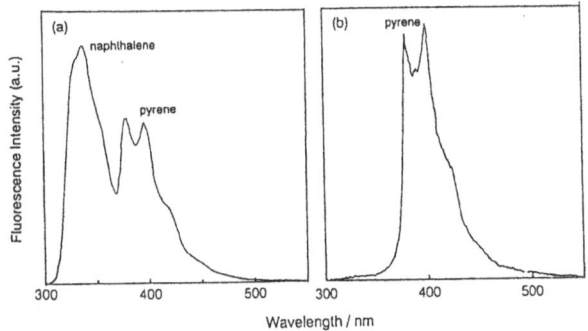

Figure 2. Fluorescence spectra for poly(A/1-Np/Py) in methanol (a) and in water (b) by excitation of the 1-Np moieties at 290 nm.

Figure 2 compares fluorescence spectra of poly(A/1-Np/Py) with selective excitation of the 1-Np chromophore (at 290 nm) in methanol and in water [16]. In methanol, in which poly(A/1-Np/Py) adopts a random coil conformation, the terpolymer exhibits both 1-Np fluorescence and Py fluorescence, the former being predominant over the latter. The random coil allows NRET from the singlet excited 1-Np to the Py sites to some extent, as can be seen from the significant intensity of the Py fluorescence (Figure 2a). In contrast, poly(A/1-Np/Py) emits only Py fluorescence in water (Figure 2b), indicating that the singlet excited energy migrates over the 1-Np clusters within the unimer and is thoroughly trapped by the Py sites. This observation suggests that, even if there are a number of cluster units in a unimer (Figure 1), all the cluster units are in contact with each other such that the 1-Np singlet energy can migrate from one cluster unit to another.

A large portion of the 1-Np fluorescence decays with a lifetime of 19 ps and there is a rise in the Py fluorescence with a time constant of 20 ps. This indicates that extremely rapid energy migration occurs throughout the 1-Np cluster and photoexcited electronic energy is rapidly trapped by the Py sites in the poly(A/1-Np/Py) unimer.

If a small amount of zinc(II) tetraphenylporphyrin (ZnTPP) is compartmentalized in the highly constraining microenvironments in the unimer in aqueous solution, the photophysical behavior of the ZnTPP can be greatly modified [19-21]. Most remarkably, the lifetime of the triplet excited ZnTPP gets extraordinarily long at ordinary or higher temperatures. This is particularly true in the Cd-containing terpolymer (Scheme 3). The triplet excited lifetimes at ordinary or higher temperatures were determined from the decays of triplet-triplet absorption in laser photolysis experiments [21]. The decay profiles for the compartmentalized ZnTPP systems were fitted with double-exponential functions, from which average triplet lifetimes were estimated. The triplet lifetime of ZnTPP in the reference copolymer (Scheme 3) is more or less 2 ms, which is about the same as the lifetime of small molecular weight ZnTPP in organic solvents. In contrast, the average lifetime in the Cd cluster is 58 ms, whereas it is 17 ms in the La cluster. The reason for the longer triplet lifetime in the Cd cluster may be due to the rigidity of the matrix where ZnTPP is compartmentalized [21].

Because the triplet lifetime of the ZnTPP species in the Cd cluster is extremely long, the terpolymer emits phosphorescence and "E-type" delayed fluorescence in aqueous solution at room temperature (Figure 3) [19,20]. This is a very unusual phenomenon for Zn-porphyrin dyes. The reference copolymer does not emit such delayed fluorescence and phosphorescence in aqueous solution at room temperature. The delayed emissions show a characteristic temperature dependence; with increasing temperature, the delayed fluorescence increases whereas the phosphorescence decreases. Analysis of the temperature dependence data indicates that the delayed fluorescence is due to the thermal activation of the triplet excited state back up to the singlet excited state.

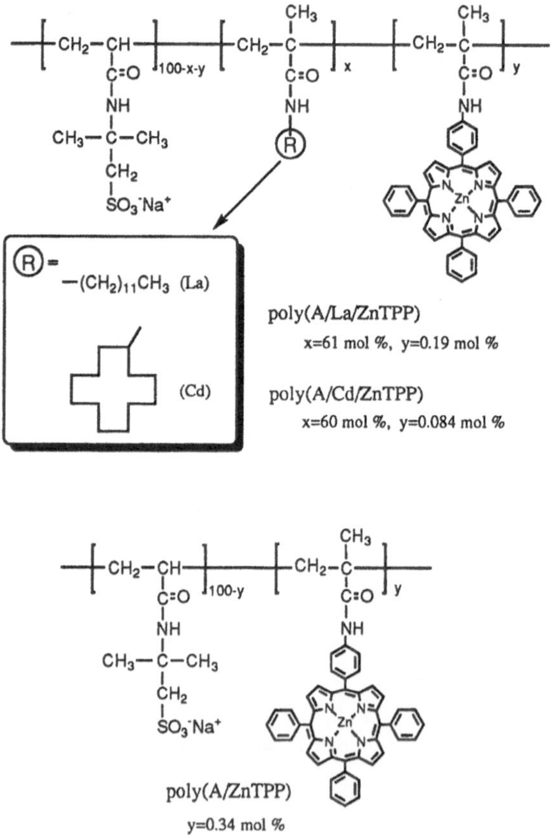

poly(A/La/ZnTPP)
x=61 mol %, y=0.19 mol %

poly(A/Cd/ZnTPP)
x=60 mol %, y=0.084 mol %

poly(A/ZnTPP)
y=0.34 mol %

Scheme 3

ZnTPP encapsulated in the La cluster also shows phosphorescence and delayed fluorescence in aqueous solution at room temperature, although their intensities are much lower than those in the Cd cluster. The phosphorescence and fluorescence spectral profiles for the ZnTPP in the Cd cluster are greatly distorted as compared with those in the La cluster; in the Cd case, phosphorescence peak is blue-shifted by about 40 nm and the intensities of the 0-0 and 0-1 fluorescence bands are reversed as compared with the spectra in the La cluster [20].

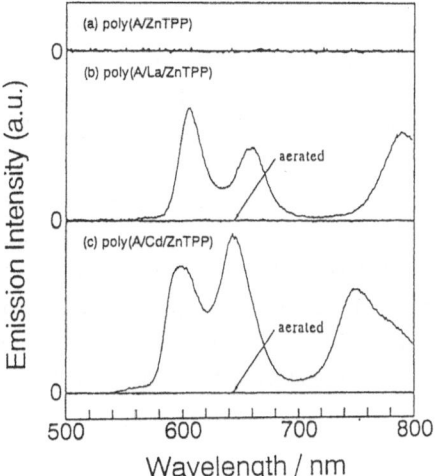

Figure 3. Delayed emission spectra in aqueous
solution at 30 °C at a delay time of 5 ms.

This difference in the microenvironment between the Cd and La cluster is also reflected in the absorption spectra; the absorption band profile in the Cd cluster is broader than that in the La cluster [20]. As described above, the Cd cluster is more rigid than the La cluster, and because of the rigid matrix the porphyrin ring may be "pinned down" to a distorted conformation in the Cd cluster which may be responsible for the unusual photophysical behavior [21].

6. REFERENCES

1 Schulz DN, Bock J, Valint Jr, PL (1994) In: Dubin P, Bock J, Davies RM, Schulz DN, Thies C (eds) *Macromolecular Complexes in Chemistry and Biology.* Springer-Verlag, Berlin Heidelberg, pp 3-13

2 Varadaraj R, Branham KD, McCormick CL, Bock J (1994) In: Dubin P, Bock J, Davies RM, Schulz DN, Thies C (eds) *Macromolecular Complexes in Chemistry and Biology.* Springer-Verlag, Berlin Heidelberg, pp 15-31

3 Chang Y, McCormick CL (1993) *Macromolecules* 26:6121

4   Morishima Y, Nomura S, Ikeda T, Seki M, Kamachi M (1995) *Macromolecules* 28:2874

5 Morishima Y, Tominaga Y, Kamachi M, Okada T, Hirata Y, Mataga N (1991) *J Phys Chem* 95:6027

6 Morishima Y, Tominaga Y, Nomura S, Kamachi M (1992) *Macromolecules* 25:861

7 Kalyanasundaram K, Thomas JK (1977) *J Am Chem Soc* 99:2039

8 Hashimoto S, Thomas JK (1985) *J Am Chem Soc* 107:4655

9 Berlman IB (1973) *Energy Transfer Parameters of Aromatic Compounds*; Academic Press, New York

10 Erdmann K, Gutsze A (1987) *Colloid Polym Sci* **265**:667

11 Raby P, Budd PM, Heatley F, Price C (1991) *J Polym Sci, Polym Phys Ed* **29**:451

12 Brereton MG, Ward IM, Boden N, Wright P (1991) *Macromolecules* **24**:2068

13 Morishima Y, Furui T, Nozakura S, Okada T, Mataga N (1989) *J Phys Chem* **93**:1643

14 Morishima Y (1992) *Adv Polym Sci* **104**:51

15 Morishima Y (1994) Trends Polym Sci **2**:31

16 Morishima Y, Tominaga Y, Nomura S, Kamachi M, Okada T (1992) *J Phys Chem* **96**:1990

17 Morishima Y, Tsuji M, Kamachi M, Hatada K (1992) *Macromolecules* **25**:4406

18 Morishima Y, Tsuji M, Seki M, Kamachi M (1993) *Macromolecules* **26**:3299

19 Aota H, Morishima Y, Kamachi M (1993) *Photochem Photobiol* **57**:989

20 Morishima Y, Saegusa K, Kamachi M (1995) *Macromolecules* **28**:1203

21 Morishima Y, Saegusa K, Kamachi M (1995) J Phys Chem **99**:4512

Optical Characterization of Ultrathin Polymer Films Prepared by the Langmuir-Blodgett Technique

S. Ito and M. Yamamoto

Division of Polymer Chemistry, Graduate School of Engineering
Kyoto University, Sakyo, Kyoto 606-01, Japan

Abstract: A variety of amphiphilic polymers are able to form a stable monolayer on the water surface, which is transferable onto a solid substrate layer by layer, yielding artificial molecular assemblies with a scale of molecular dimension. This report describes the structural characterization of ultrathin polymer films prepared by the Langmuir-Blodgett technique. The optical methods we employed sensitively detect the nano-structure of mono- and multi-layered polymer films, such as layer thickness, distance, thermally induced relaxation, and also morphology of the surface monolayer, showing the specific features of polymer chains spread on a two-dimensional plane.

INTRODUCTION

The ultrathin polymer film prepared by the Langmuir- Blodgett(LB) technique is a fascinating material with many applications[1]. Due to the thinness of each layer, the designed structure in a molecular dimension allows one to give the polymer film various new functions[2]. However, we are often faced with difficulties in characterizing the structure because the specimen examined has a thickness of only a few nanometers, that is, we absolutely need highly sensitive techniques which must have spatial resolution in the range of molecular scale and must be able to detect weak signals from a monolayer. Recently, we have applied various optical methods including fluorescence spectroscopy[3-8], surface plasmon spectroscopy[9] and Brewster angle microscopy[10,11] to probe the nano-structure of polymer LB films. Here, we demonstrate some applications of these optical methods for studying ultrathin polymer films.

CHARACTERISTICS OF POLYMER LB FILMS

Since the pioneering works of Kuhn and his coworkers, extensive investigations have been made on LB films of long chain fatty acids to design and construct functional thin films[12]. Recently, the LB technique was found to be applicable also to some preformed polymers which form a stable monolayer at the air-water interface under an appropriate surface pressure[2]. Polymers have many advantages as the LB materials. For example, their chemical structures are easily modified by chemical reactions introducing various functional units with covalent bonds. The polymer LB films obtained have some remarkable properties, such as mechanical stability,

M. Kamachi · A. Nakamura (Eds)
New Macromolecular Architecture and Functions
Proceedings of the OUMS '95 Toyonaka, Osaka, Japan, 2-5 June, 1995
© Springer-Verlag Berlin Heidelberg 1996

amorphous character, and fewer pinholes, which are never realized in the conventional fatty acid LB films.

From the view point of photo-functional polymer films, they have several advantages in addition to the features mentioned above. One is the thinness of each layer. The polymer monolayer is stabilized both by the cohesive force of the side chains and by the chemical bonds of the main chain, therefore the hydrophobic side chain length is much shorter than the alkyl chain of fatty acid LB films. This results in the thinness of the individual layer. Usually, the thickness is only about 1 nm; the thinness is a very important character to control interlayer reactions such as energy transfer and electron transfer between layers because these transfer processes have a reaction radius of ca. 1 - 2 nm. The other is the uniform distribution of chromophores attached to the polymer chain[13]. The amorphous character of the polymer film prevents the chromophores from forming the ground state dimers and aggregates, and keeps the chromophores in a uniform distribution in a two-dimensional plane. From the application standpoint of LB films, the mechanical and thermal stabilities are critical. Polymer LB films have been expected to be more stable than the conventional fatty acid LB films. Although the polymeric LB films have so many favorable features, a drawback is the poor regularity of the structure in the atomic level due to the amorphous properties of polymeric materials. However, in place of it, polymer LB films have homogeneity at the molecular level, which is of great importance for most applications.

In the present study, we employed poly(vinyl alkanal acetal) (abbreviated as PVO for octanal acetal, PVD for dodecanal acetal) as the representatives of LB polymers[14]. As shown in Scheme 1, PVO was prepared from poly(vinyl alcohol) (PVA) by acetalization with octanal. The mixture of these chemicals was kept at 40 °C overnight in a chloroform solution containing a drop of hydrochloric acid. A dilute solution of the obtained polymer was spread on the water surface by the usual LB method. It forms a stable monolayer and then was transferred onto a solid substrate layer by layer with a good transfer ratio. Figure 1 shows the thickness of the LB films measured by ellipsometry. The thickness is plotted against the number of depositions. From the slope of the straight line, the thickness of one layer can be evaluated to be only about 1 nm. Therefore, the number of layers is nearly equal to the film thickness presented by the nm unit.

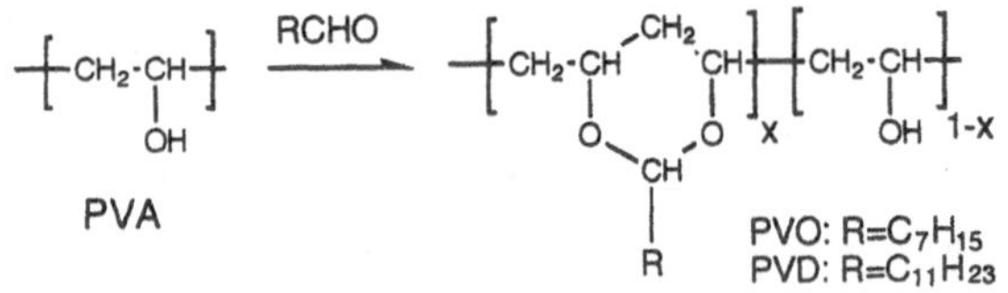

Scheme 1

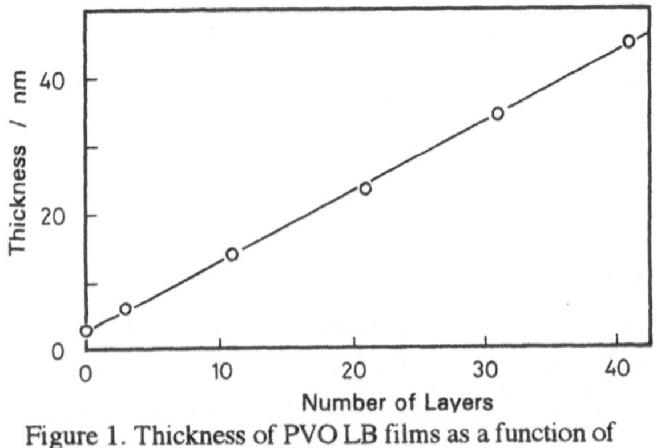

PVO: R=C$_7$H$_{15}$
PVD: R=C$_{11}$H$_{23}$

Figure 1. Thickness of PVO LB films as a function of number of layers.

BREWSTER ANGLE MICROSCOPY

From the standpoint of morphology of a monolayer at the air-water interface, a long polymer chain must be confined in a 2-dimensional form on the water surface. Therefore, the direct observation of monolayer at the air-water interface is highly desirable to build a nano-structure. Generally speaking, however, it is difficult to see a monolayer on the water surface because it is too thin to obtain an image with sufficient optical contrast between the areas of the monolayer and the water surface. In 1991, two European groups developed a Brewster angle microscope (BAM), which depicts clearly a monolayer with sufficient contrast[10,11]. When a light beam illuminates the air-water interface at the Brewster angle, θ_B which is defined by the ratio of refractive indices of two media as follows,

$$\theta_B = \tan^{-1}(n_{water} / n_{air}), \quad (1)$$

only s-polarized light is reflected with no reflection at all for the p-light. Figure 2 shows the reflectivity plot against the angle

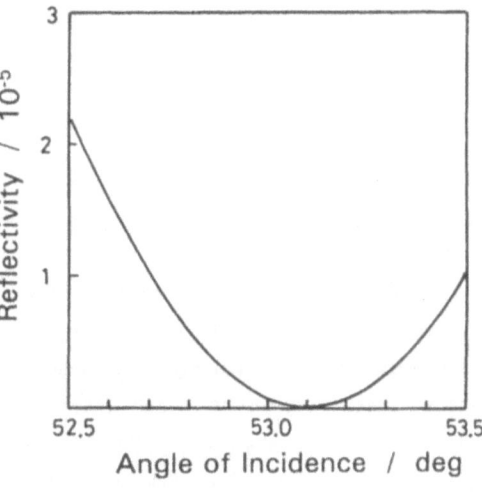

Figure 2. Reflectivity for p-polarized light at the air/water interface: $n_{air} = 1.00$, $n_{water} = 1.33$.

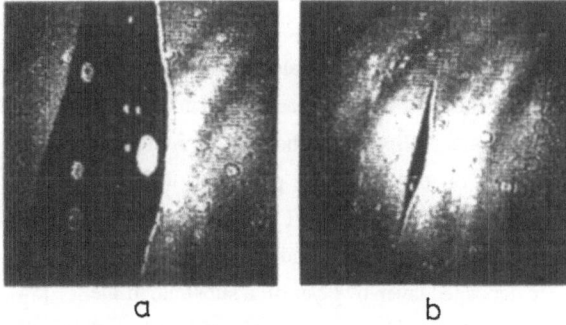

Figure 3. BAM images of PVD monolayer at the air/water interface. The width of each picture is 1 mm.

of incidence. As for the air-water interface, the Brewster angle is $53.1°$ for the He-Ne laser light of 632.8 nm, and at this angle of incidence the reflectivity of p-light becomes practically zero. However, when the water surface is covered with a thin polymer film, this angle is no more the definite Brewster angle for the covered area, therefore, very weak but detectable intensity of light can be obtained from the area of the monolayer. By the microscopic observation of the surface just at θ_B, the surface morphology of the polymer monolayer can be visualized as a bright image of the monolayer in the dark background of the water surface.

Figure 3 is an example of a BAM picture of a poly(vinyl dodecanal acetal) (PVD) monolayer. This polymer has unreacted hydroxy units in the main chain, which give the polymer chain an amphiphilic character and also flexibility. The pictures (a and b) are two successive shots taken at a 1 min interval, under compression of the monolayer. The black area presents the bare water surface and the white one is the monolayer domain. The border of each domain forms a very smooth line and changes the shape with time, showing a very viscous but liquid-like character.

130

With the compression of the monolayer, two domains coalesce each other and eventually form a uniform monolayer. It is quite clear from these pictures that the polymer monolayer behaves like a viscous liquid or polymer melt on the water surface. This is the reason why this polymer is easily transferred onto various solid substrates, yielding a very thin and uniform coating without serious defects such as cracks and pinholes.

FLUORESCENCE METHODS

Highly sensitive techniques are indispensable to probe the nano-structure of ultrathin multilayers. We have employed two kinds of optical spectroscopies for this purpose: fluorescence spectroscopy and surface plasmon spectroscopy. As for the fluorescence spectroscopy, the energy transfer phenomenon is utilized as a nano-ruler of the layer distance[3]. The chromophore phenanthrene (P) unit as an energy donor on excitation, or anthracene (A) as an energy acceptor, is covalently linked to the polymer chain by the same acetal reaction as the alkanal unit. Figure 4 shows the molecular structures of these labeled polymers with the compositions listed in Table 1. These fluorescent probe labeled monolayers and non-labeled layers were deposited layer by layer on a substrate in the sequence shown in Figure 5. The donor P and acceptor A layers were separated by an appropriate number of spacing layers in order to control the energy transfer efficiency between the donor and acceptor layers. These samples are abbreviated as PnA, where n is the number of spacing layers.

Figure 6 shows fluorescence spectra of PnA samples observed when the donor P is selectively excited by UV light. The spectra consist of P emission around 350 nm and A emission around 400 - 450 nm which is emitted by the efficient energy transfer from the excited P to A. With the decrease in the number of spacing layers, the efficiency of energy transfer becomes higher and higher, and in the case of P0A, most energy on P

Table 1. Compositions of synthesized PVO samples for energy transfer measurements

Sample	x (%)	y (%)	1-x-y (%)
PVO	0	73	27
PVO-P	12	57	31
PVO-A	7	55	38

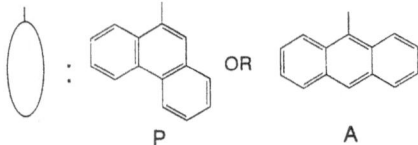

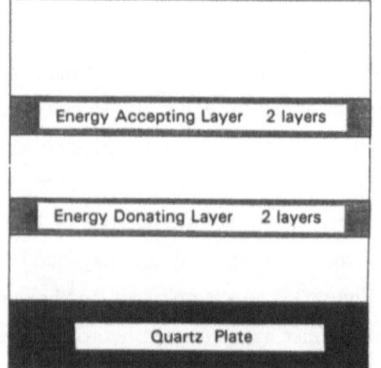

Figure 4. PVO samples bearing phenanthrene (PVO-P) and anthracene (PVO-A) chromophores.

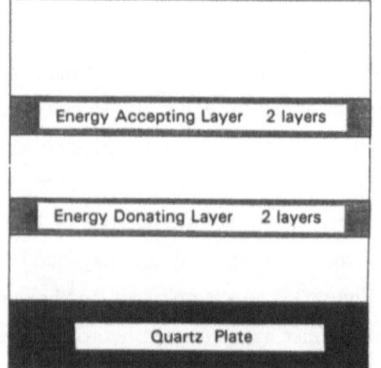

Figure 5. Schematic illustration of the multilayer structure of LB film for the energy transfer measurement.

is transferred to the A layer, yielding predominant emission from A. Thus, we are able to control one of the fundamental photo-processes, namely, the energy transfer process by the artificial nano-structure.

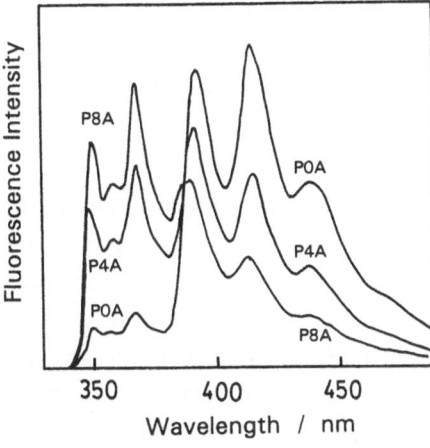

Figure 6. Fluorescence spectra of PnA LB films. The P unit is selectively excited at 298 nm.

The layer distance, in more detail, the distribution of distance of separation can be determined by measuring the efficiency of energy transfer between the P and A chromophores[7]. The energy transfer rate constant from a certain donor i to an acceptor j is given by

$$k_{Tij} = (1/\tau_0) (R_0 / r_{ij})^6 \qquad (2)$$

where r_{ij} is the distance of separation between i and j. R_0 is the critical transfer distance determined by the pair of donor and acceptor chromophores, at which the rate of transfer is equal to the donor's intrinsic deactivation rate: $1/\tau_0$[15]. These parameters for the P and A pair are obtained experimentally as $R_0 = 2.12$ nm and $\tau_0 = 43$ ns. The energy transfer takes place to all of acceptors, therefore the rate will be given by the sum of eq. 2.

$$k_{Ti} = (1/\tau_0) \Sigma_j (R_0 / r_{ij})^6 \qquad (3)$$

The probability $p_i(t)$ that the excited donor i survives at a time t after the excitation is calculated as follows,

$$p_i(t) = \exp (-t/\tau_0 - k_{Ti}t) \qquad (4)$$

Therefore, the observed p(t) for the system is given by the average of $p_i(t)$ over all donors,

$$p(t) = (1/n_D) \Sigma_i p_i(t) \qquad (5)$$

here n_D is the number of donors in the system, and the p(t) corresponds to the fluorescence decay curve observed. Under the steady state excitation, the integration of eq. 5 gives the fluorescence quantum yield q_D of the donor,

$$q_D = k_f \int p(t)\, dt \qquad (6)$$

where k_f is the rate of the fluorescence emission of the donor. Therefore, the energy transfer efficiency q_T is given by

$$q_T = (q_{D0} - q_D) / q_{D0} \qquad (7)$$

where q_{D0} is the quantum efficiency of donor emission given by

$$q_{D0} = k_f \tau_0. \qquad (8)$$

All of these eqs. 2-7 are numerically calculated for a given coordinate of D and A chromophores, and both the transient decay curve p(t) and the energy transfer efficiency q_T can be compared with the experimental values.

To fit the experimental data with the theoretical predictions, we have to introduce a Gaussian distribution of chromophores in the direction normal to the film plane as shown in Figure 7. The distribution is characterized by the standard deviation σ and the mean displacement x_0.

$$c(x) = [c_0 / \sigma(2\pi)^{1/2}] \exp [-(x-x_0)^2 / 2\sigma^2]$$

$$(9)$$

where $c(x)$ is the concentration of chromophores at the displacement x, and c_0 is related to the total number of chromophores placed in the system.

Figure 8 shows an example of the decay curve fitting for the P4A film. A good fit could be obtained by assuming a distribution of chromophores with a standard deviation of $\sigma = 1.35$ nm. The solid lines in Figure 7 represent the distribution of chromophores, and the broken lines are the positions at which the each layer had been deposited. Since all layers were built up as a Y-type film, the chromophoric layers consist of two layers with a thickness of 2 nm at which the standard deviation is not so large initially. However, if the LB film is placed under a high temperature, disordering proceeds rapidly, resulting in an increase of the σ-value with the elapse of time. These findings indicate that the polymer LB films have some extent of disordering. Although macroscopic measurements such as UV absorption and ellipsometry show no change before and after the aging, the fluorescence method based on the interlayer energy transfer is able to detect sensitively the alteration of film structure in the molecular dimension.

PHOTOCHEMICAL STABILIZATION OF THE LAYER STRUCTURE

The structural relaxation of polymer LB films could be probed by the fluorescence method. An effective strategy to improve the thermal stability is the use of photochemical crosslinking after the LB deposition[9]. The cinnamoyl group, known as the functional group of a classical photoresist, was introduced into PVO in which the remaining hydroxy units were replaced by cinnamoyl units as shown in Figure 9. Thus, we obtained a

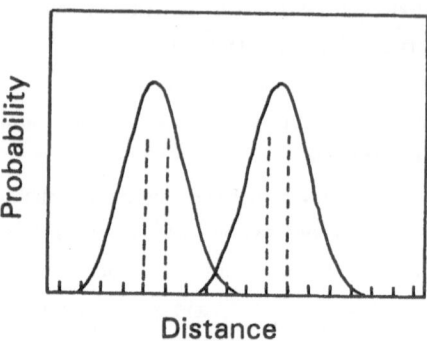

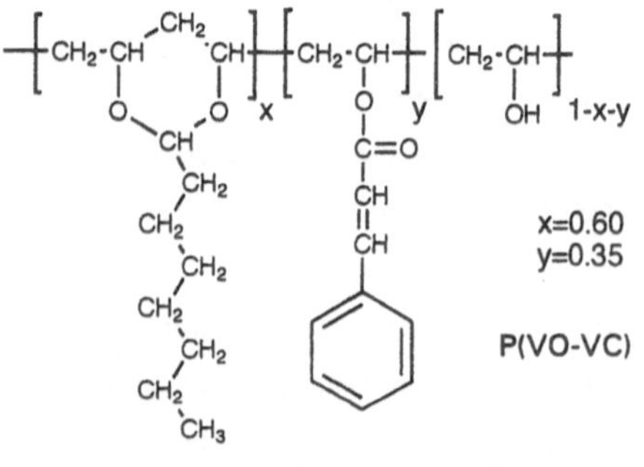

Figure 7. Distribution of chromophores in a P4A LB film. The solid lines show a Gaussian distribution with a standard deviation of 1.35 nm.

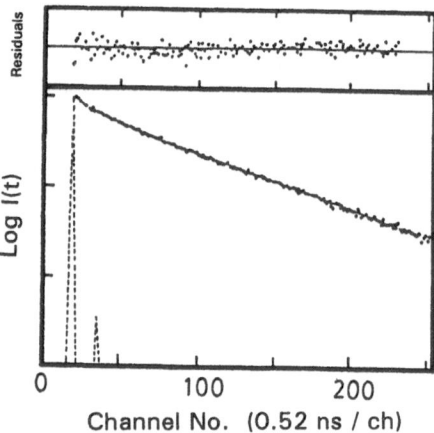

Figure 8. Decay curve fitting, assuming a Gaussian distribution for P4A.

Figure 9. Chemical structure of the cinnamate polymer, P(VO-VC).

photo-reactive polymer P(VO-VC) which contains about 35 % cinnamoyl units.

The monolayer of P(VO-VC) is also transferable onto a quartz plate and we prepared a 20-layer cinnamate LB film for the spectroscopic measurements. The absorption band of cinnamoyl unit appears around 280 nm. This band was selectively excited by monochromatic light from a high pressure Hg lamp, and the photoreaction was monitored by the decrease of the absorbance. As shown in Figure 10, UV irradiation caused a rapid decrease of absorbance due to efficient photo-dimerization of cinnamoyl units. The reaction was so fast that the decrease of absorbance leveled off within 5 min of irradiation. Judging from this value, ca. one third of the cinnamoyl units can react with each other. The stabilization effect of photo-crosslinking was confirmed by examining the solubility of the irradiated film into a good solvent, dichloromethane. After the photoreaction, the sample film was dipped into the solvent, and the remaining fraction of cinnamoyl absorbance was measured as a function of dipping time (Figure 11). One minute irradiation made the LB films completely insoluble, while the absorbance of the control sample without irradiation rapidly decreased within a few minutes.

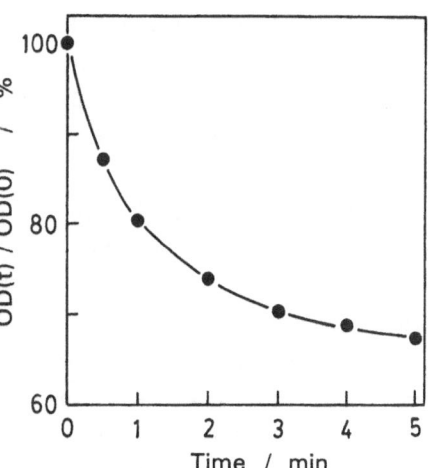

Figure 10. Decrease of cinnamoyl absorbance as a function of UV irradiation time.

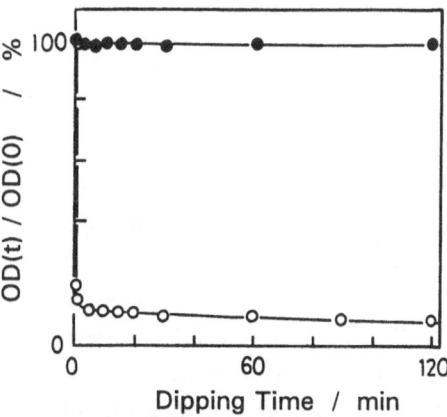

Figure 11. Solubility of the UV irradiated P(VO-VC) 20-layer film against dipping time in dichloromethane:irradiation for (●) 1 min, (○) 0 min.

SURFACE PLASMON SPECTROSCOPY

The surface plasmon resonance is very sensitive to alteration of the thickness and refractive indices of the coating. Let us suppose that the bottom of the prism is covered with a thin silver layer, and a laser beam hits the prism at an incident angle, θ_{ex} (see Figure 12). If the angle is larger than the critical angle of total reflection, θ_c, most of the light is reflected from the bottom of the prism, and a strong intensity of light is recorded by the detector. However, at a particular

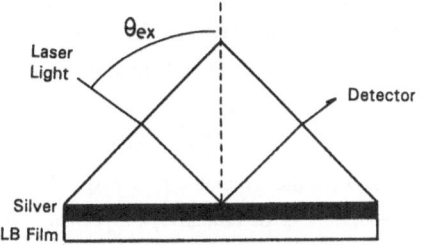

Figure 12. Setup for the excitation of surface plasmon in Kretchmann geometry.

angle, θ_{sp}, in the present case 42.0° for the silver layer, the intensity becomes zero, because the electromagnetic wave of light is coupled with the plasma oscillation of electrons at the silver surface. This resonance angle is very sensitive to the change of optical properties of dielectric coating.

We applied the surface plasmon spectroscopy to characterize the thickness changes of the cinnamoyl LB film described above[9]. Figure 13 depicts an example of reflectivity plots against the angle of incidence. First, a sharp dip appeared at 42.0° for a bare silver: curve (a). On the top of the silver, 4 layers of PVO and 8 layers of cinnamoyl LB films were deposited, resulting in a resonance shift to 43.8°: curve (b). And then the sample was irradiated with a photo-mask and dipped into dichloromethane. The resonance for the non-irradiated part returned to 42.0°, showing that the LB films were completely removed from the surface by the dipping process. The irradiated part also showed a slight shift of resonance to lower angles, probably a part of

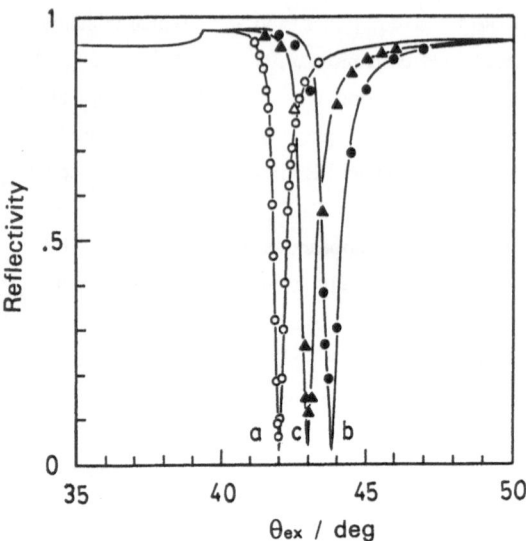

Figure 13. Reflectivity curves observed for (a) bare silver, (b) a LB film composed of 4-layer PVO and 8-layer P(VO-VC) on silver, (c) the film irradiated and immersed in dichloromethane.

Table 2. Thicknesses (d) and resonance angles (θ_{sp}) obtained by surface plasmon measurement

Sample	d (nm)	θ_{sp} (deg)
bare silver	56.0	42.0
PVO(4 layers)	4.1	43.8
P(VO-VC) (8 layers)	7.6	
irradiated part	7.6	43.0
non-irradiated part	0	42.0

PVO layers dissolved into the solvent. The solid lines in this figure present the calculated curves by Fresnel equation using the thicknesses listed in Table 2. Table 2 shows the film thickness of each coating: 4.1 nm for the precoated PVO layers and 7.6 nm for the cinnamoyl layers (P(VO-VC)). Therefore, the total ca. 12 nm for the whole LB film. For the non-irradiated portion after the development, the resonance signal became exactly the same as that for the original bare silver. This fact indicates that the thickness returned to 0 nm. However, the thickness for the irradiated part was 7.6 nm after the solvent treatment. Since this value is equal to the thickness of cinnamoyl layer, the precoating PVO layers seem to be dissolved from the surface. It is notable that the underlying layers were selectively dissolved. However, a similar case has been seen previously: polyimide LB film. During the imidization process of the LB film of poly(amic acid) alkylamine salt, the LB film loses long alkyl chains by the solvent treatment. Eventually, the film is converted to ultrathin multilayer films of polyimide with each thickness of only 0.5 nm. As demonstrated by the photoreactive LB film, the surface plasmon technique allows us to evaluate precisely the film thickness and to observe chemical processes taking place in the nano-films.

CONCLUSION

We described some specific features of ultrathin polymer films prepared by the LB technique. The obtained nano-structure could be characterized and visualized by various optical spectroscopies. Generally speaking, the optical method has many advantages as a probing technique: non-contact and in situ observation, microscopic approach and visualization, and extremely high sensitivity and time-resolution. These are particularly favorable for the thin film research, and the nice combination of the nano-materials and optical techniques enables one to clarify the fascinating characters and functions provided by the fabricated structures of ultrathin polymer films.

ACKNOWLEDGMENT

We are particularly grateful to our colleagues, S. Ohmori, T. Hayashi, T. Okuyama, M. Mabuchi, and N. Sato for their enthusiasm during their thesis work. We also thank Prof. W. Knoll of Max-Planck-Institute, Germany, for his kind guidance on surface plasmon spectroscopy. This work was partly supported by a Grant-in-Aid for Scientific Research (No. 06651050) and by a Grant-in-Aid for Scientific Research on Priority Areas (No. 06239107) from the Ministry of Education, Science and Culture of Japan.

References:

[1] Roberts GG (1985) Adv. Phys., 34: 475

[2] Tredgold RH (1987) Thin Solid Films, 152: 223

[3] Ohmori S, Ito S, Yamamoto M (1991) Macromolecules, 24: 2377

[4] Ito S, Kanno K, Ohmori S, Onogi Y, Yamamoto M (1991) Macromolecules, 24: 659

[5] Ueno T, Ito S, Ohmori S, Onogi Y, Yamamoto M (1992) Macromolecules, 25: 7150

[6] Ito S, Ueno T, Yamamoto M (1992) Thin Solid Films, 211: 614

[7] Yamamoto M, Kawano K, Okuyama T, Hayashi T, Ito S (1994) Proc. Japan Acad., 70, Ser. B: 121

[8] Hayashi T, Okuyama T, Ito S, Yamamoto M (1994) Macromolecules, 27: 2270

[9] Hayashi T, Mabuchi M, Mitsuishi M, Ito S, Yamamoto M, Knoll W (1995) Macromolecules, 28: 2537

[10] Hönig D, Möbius D (1991) J. Phys. Chem., 95: 4590

[11] Henon S, Meunier J (1991) Rev. Sci. Instrum., 62: 936

[12]Ulman A (1991) An Introduction to Ultrathin Organic Films from Langmuir-Blodgett to Self-Assembly, Academic Press, San Diego

[13]Ohmori S, Ito S, Yamamoto M (1990) Macromolecules, 23: 4047

[14]Watanabe M, Kosaka Y, Oguchi K, Sanui K, Ogata N (1988) Macromolecules, 21: 2997

[15]Förster Th (1949) Z. Naturforsch, 4a: 321

Excited State Electron Transfer from Polyelectrolyte-Bound Chromophores

S.E. Webber

Department of Chemistry and Biochemistry and Center for Polymer Research
The University of Texas at Austin, Austin, Texas 78712 (USA)

INTRODUCTION AND OVERVIEW[1]

The transduction of light into chemical potential has been actively studied via a variety of mechanisms. Perhaps the most actively pursued approach is via photoredox chemistry, which may be generically represented by the following:

$$D + h\nu \rightarrow {}^{2S+1}D* \tag{1}$$

$$^{2S+1}D* \xrightarrow{k_0} D + heat \tag{2a}$$

$$^{2S+1}D* + A \xrightarrow{k_q} {}^{2S+1}[D^{+\bullet}, A^{-\bullet}] \tag{2b}$$

$$^{2S+1}[D^{+\bullet}, A^{-\bullet}] \xrightarrow{k_{cs}} D^{+\bullet} + A^{-\bullet} \tag{3a}$$

$$\xrightarrow{k_b} D + A + heat \tag{3b}$$

with a quantum yield of charge separation given by

$$\phi_{cs} = \frac{k_{cs}}{k_{cs} + k_b}. \tag{4}$$

The fraction of excited states quenched in step (2) is given by

$$f_q = \frac{k_q[A]}{k_0 + k_q[A]}. \tag{5}$$

In step (2b) a geminate pair of ion-radicals is formed, and in (3a) they diffuse apart to form uncorrelated ion pairs that can participate in some useful redox chemistry. Useful chemistry must compete with the bulk recombination process

$$D^{+\bullet} + A^{-\bullet} \xrightarrow{k_{rec}} D + A + heat \tag{6}$$

where k_{rec} is a typical second order rate constant. If $D^{+\bullet}$ and $A^{-\bullet}$ remain in the same bulk phase then this process is expected to be facile. Step (6) should be distinguished from the geminate recombination in step (3b).

Optimization of this transduction requires the following:

(1) Efficient production of the desired excited state $^{2S+1}D*$ at the wavelength(s) of interest, which is determined by the extinction coefficient, concentration of D, formation yield and lifetime of $^{2S+1}D*$. This obviously can be controlled only by the choice of chromophore, and to a lesser extent, the solvent and/or other environmental features.

(2) The excited state $^{2S+1}D*$ must be efficiently quenched, which depends on the excited state properties (through k_0 and the excited state redox potential) and the quencher (through the product $k_q[A]$), as expressed in eq.(5). Of course, the concentration [A] represents the local concentration around the D moiety in the case of heterogeneous systems, such as discussed herein.

M. Kamachi · A. Nakamura (Eds)
New Macromolecular Architecture and Functions
Proceedings of the OUMS '95 Toyonaka, Osaka, Japan, 2-5 June, 1995
© Springer-Verlag Berlin Heidelberg 1996

(3) The quantum yield of charge separation (ϕ_{cs} in eq.(4)) is perhaps the most critical bottleneck in light energy transduction, and is the step that has played a central role in our research. ϕ_{cs} depends on many properties such as the strength of interaction between $^{2S+1}D^*$ and A, the energetics of the electron-transfer step, and solvation of the $^{2S+1}[D^{+\bullet}, A^{-\bullet}]$ geminate pair or $D^{+\bullet} + A^{-\bullet}$ bulk solvent pair.

(4) A high ϕ_{cs} is not of any practical utility unless the $D^{+\bullet}$ and $A^{-\bullet}$ species live long enough to carry out useful chemistry. In our work (and that of many others) the species persist for long periods of time by chemical standards (i.e. milliseconds), but this is often because of the low overall bulk concentration of the ion-pairs, since k_{rec} (eq.(6)) is usually diffusion limited. Therefore if one achieved a high density of ion-pairs it would be difficult to compete with the wasteful back reactions. Many investigators have approached this problem by phase separation of the ion-pair via membranes, vesicles, or surfactant micelles.[2] In our research we have used hydrophobic polymer-water interfaces to accomplish this goal. This is illustrated in Scheme 1, which will be discussed in more detail in later sections.

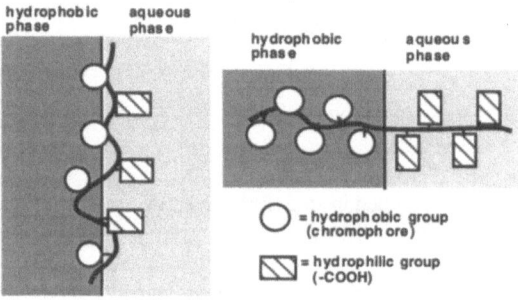

Scheme 1

It is appropriate to review how this research has evolved over time. While a variety of aromatic chromophores has been studied, we summarize the results for various derivatives of anthracene, which have been most broadly investigated by us and which illustrates the major results.

Our first work used polymethacrylic acid (PMA) with a small mole fraction of copolymerized anthracene derivative (PMA-A). It was found that for the singlet state charge separation could be observed only for rather low pH (< 3), where PMA is known to be "hypercoiled". This led to the idea that some sort of "hydrophobic protection" of the chromophore is required for effective singlet state charge separation. While the precise nature of "hydrophobic protection" remains vague, we have always found that if there is appreciable static quenching ϕ_{cs} is low. For the triplet state charge separation is observed at high and low pH, consistent with the idea that geminate recombination to the ground state is spin-forbidden (see eq. (3b)). At high pH the negative charge density of the polyelectrolyte further encourages charge separation.

The next set of experiments was to adsorb these PMA-A polymers onto polystyrene (PS) latex particles (referred to as microspheres, abbreviated μS). Surprisingly, ca. 1 mol % of a hydrophobic species greatly enhances this adsorption. The charge separation in the singlet state was appreciable even at high pH, which was interpreted as the result of adsorption of the chromophore or the polystyrene surface, thereby providing the required hydrophobic protection. ϕ_{cs} for the triplet state was comparable to homogenous solution.

The PMA-A polymer has a random distribution of chromophores, and the next approach was to place the chromophore precisely at the end of a water soluble polymer. Polyethylene oxide was used (PEO-A), and this polymer also adsorbed onto PS latexes, with good singlet state charge separation. However this polymer also exhibited excellent singlet state charge separation in homogenous solution. Thus it would seem that PEO mimics PMA at low pH with respect to hydrophobic protection (PEO is a well-known solubilizing agent for hydrophobic materials).

Our most recent work involves diblock polymers of polystyrene and polymethacrylic acid with either a single anthracene or a short anthracene block inserted between the blocks (denoted PS-A-PMA). These materials spontaneously self-assemble into micelles. It is expected that the anthracene moieties are located near the PS-H_2O interface, seems as though it should be an ideal morphology for hydrophobic protection. However it would seem that the hydrophobic protection is too good, such that it is difficult to quench these fluorophores. The pH effects on quenching are much smaller than for linear polymers and the charge separation quantum yield is disappointing (ca. 0.05, see Table 1). However, the lifetime of the ion pair is remarkable, longer than we can measure with the standard transient absorption apparatus (> 10 ms) and in fact steady-state irradiation can be used to build up a significant concentration of SPV$^{-\bullet}$ anion radical (see structure **1** later). Furthermore the SPV$^{-\bullet}$ is quenched by O_2 to a much smaller extent than in homogenous solution, presumably because it resides in the micelle corona region. This implies that the PMA corona is not deprotonated near the interface, even at pH 9. A great deal more remains to be done with these materials [3]. In all our current work we are emphasizing the interface between a hydrophobic polymer and water. The reasons for this may be summarized:

(1) A solar energy scheme based on thin polymer films in contact with an aqueous solution of reactant would have tremendous advantages for device construction and a very large range of polymers can be explored.

(2) We believe that the photophysical experiments and classical characterization we will carry out will provide unique insights to the behavior of polymers at solid and liquid interfaces, which is a very important scientific and technological problem in its own right.

All the information discussed above is compiled in Table 1 which provides a broad overview of our research on photoredox processes in polymers for the specific case of anthracene chromophores.

Table 1: Charge separation yields (ϕ_{CS}) ($\pm10\%$ est. error)

system	pH	ϕ_{CS}	comments
PMA-^{1}A*(aq.soln)[a]	2.8	0.21	original polyelectrolyte work, illustrates effect of "hydrophobic
ditto	11	ca.0.05	protection"
PMA-^{3}A*(aq.soln)[b]	4	0.30	illustrates that the effect of spin on the back reaction can overcome
ditto	11	0.89	lack of "hydrophobic protection"
PMA-^{1}A*(μS)[c]	4	0.64	illustrates that the μS provides "hydrophobic protection" even at
ditto	11	0.39	high pH
PMA-^{3}A*(μS)[c]	4	0.24	combined "hydrophobic protection"
ditto	11	0.95	and effect of spin
PEO-^{1}A*(μS)[d]	NA	0.32	another example of using latexes, only
PEO-^{3}A*(μS)[d]		0.60	now with end-tagged polymer
PEO-^{1}A*(aq. soln) [d]		0.29	PEO as hydrophobic protector in homogeneous solution
PS-^{1}A*-PMA(micelle)[e]	4-11	0.03-0.10[+]	first example of using polymer micelle as photoredox medium; only slight pH dependence; exceptionally long ion-pair lifetime

(a.) Stramel, R.D. et al. J. Phys. Chem. **1989**, *93*, 1928 ;(b).Hsiao, J.S.; Webber, S.E. J. Phys. Chem. **1992**, *96*, 2892; (c). Hsiao, J.S.; Webber, S.E. J. Phys. Chem. **1993**, *97*, 8296; (d) Hsiao, J.S.; Eckert, A.R.; Webber, S.E. J. Phys. Chem. **1994**, *98*, 12032; (e) Eckert, A. R.; T. J. Martin; Webber, S.E. (manuscript in preparation). Yield data preliminary and represents a lower limit.

[+] precise yield depends on anthracene derivative.

PHOTOPHYSICAL METHODOLOGY

Quenchers/Electron Acceptors: In all our work we have used members of the viologen family as electron acceptors, either the dicationic methyl viologen or the zwitterionic SPV, **1**. In general we prefer the latter species because it has a reduced tendency to form charge transfer complexes with aromatic chromophores. These species are well understood as long-lived electron acceptors and can be extensively modified chemically. We have also used simple ions such as Tl$^+$ and Cu^{+2} or neutral water-soluble species such as CH_3NO_2, for which no long-term charge separation is expected, in order to assess the exposure of the chromophore to the aqueous phase.

$^-$O$_3$S $\diagup\diagup$ $\overset{+}{N}$ $\diagup\diagup$ $\overset{+}{N}$ $\diagup\diagup$ SO$_3$$^-$

SPV

1

Quenching Studies: For a given system and pH condition the fluorescence (singlet state) quenching is characterized. For a general polyanion (e.g. poly(styrene sulfonate) or poly(methacrylic acid at high pH) electrostatic attractions for the viologen (even for neutral SPV, presumably via dipolar attraction) can result in very effective quenching, which can be orders of magnitude larger than diffusion controlled. For triplet state studies much lower quencher concentrations are required because of the long triplet state lifetime. It is necessary to establish that little or no singlet state quenching is occurring under the conditions of the triplet-state experiment. The magnitude and shape of the Stern-Volmer curve for steady state (I_0/I) or time-dependent ($<\tau_0>/<\tau>$, where the symbol $<\tau>$ denotes the average lifetime of the fluorescence decay) provide considerable insight as to the exposure of the chromophore to the aqueous phase and the nature of the chromophore-quencher interaction (e.g. compare dynamic or static quenching [4]).

Transient Absorption Spectroscopy: The transient spectroscopy of the polymer is characterized in the absence of quencher to establish the threshold for photoionization. We have found these systems to be very easily photoionized, presumably because of the aqueous environment and high local charge density. Photoionization must be eliminated or minimized because the products of photoionization in the presence of a quencher are the same as the desired photoredox process (see eq. (1-3)):

$$D + 2h\nu \rightarrow D^{+\bullet} + e^-(aq) \qquad (7)$$

$$e^-(aq) + A \rightarrow A^{-\bullet} \qquad (8)$$

While this facile photoionization could be useful under certain circumstances (e.g. laser photo processing), it is not important under low fluxes typical of solar radiation, and could lead to an overestimate of the charge separation yield. We have also found that the extinction coefficients for some species (e.g. ^{3}pyrene*) are modified in the aqueous environment and that to obtain valid quantum yields the new values had to be determined.[5] It is not always easy to obtain accurate transient absorption spectra for micelle or latex systems because of light scattering.

Yield per Quenching Event: The yield of ion pairs per quenching event is determined from transient absorption spectroscopy and knowledge of the quenching kinetics. In the case of the triplet state the quenching kinetics are also determined by transient spectroscopy because the triplet cannot be observed by emission in most fluid solutions. The lifetime of the ion-pair is also determined by transient absorption. In all cases that we have found ϕ_{cs} to be appreciably above zero the ion-pair persisted into the millisecond region. This is in sharp contrast to most studies in organic solvents in which ion-pair states persist for tens of μs. This is almost certainly because of the ability of water to solvate one or both ions. No attempt has been made to characterize the turnover number of the aromatic chromophores we have studied. Qualitatively we can state that long term laser excitation permanently degrades these materials, although we have verified that it is not necessary to use flow cells to obtain valid data. Clearly this class of chromophores could not be used in a practical solar transduction device, as is already obvious from their absorption spectrum (λ_{max} = 360-400 nm). However we believe that by studying a relatively narrow class of chromophores in a variety of organized environments that we can best elucidate the role of the environment. However it has to be kept in mind that a very different class of chromophores (porphyrins for example) might respond quite differently to these interfacial environments.

CHARACTERIZATION OF MATERIALS

Polymer Characterization: Both block and graft polymers must be characterized with respect to composition, usually with UV spectroscopy and NMR, and molecular weight. The latter is usually carried out with gel permeation chromatography (GPC) which gives only a qualitative measure of the molecular weight distribution unless the molecular weight standards available are identical to the polymer being examined. In the case of block polymers a sample of the first block is removed from the anionic polymerization system for separate characterization. Hence the relative lengths of the two blocks are very well known. The use of a diode-array UV-vis absorption detector is extremely useful in this work since we insert chromphores into the polymer chain.

We have found it to be advantageous to determine the modifications of the interfacial surface tension for H_2O-organic immiscible pairs for these polymers, which we can do by the "spinning drop method".[6] However if the polymers are strongly aggregating the interpretation of the Gibbs equation

$$\frac{-1}{RT}\frac{\partial \gamma}{\partial \ln(c)} = \Gamma \tag{9}$$

is more complex.[7] In any case this measurement determines if a given polymer is surface active.

Micelle Characterization: The properties of these polymers in mixed H_2O-organic solvents must also be determined. Amphiphilic polymers have a very strong tendency to form aggregates (large polydispersity) and/or micelles with a narrow distribution of sizes. The exact procedure for changing a mixed solvent from mixed organic and water to pure water can greatly change the state of aggregation and this topic alone remains an active area of research. These properties can be characterized via quasi-elastic light scattering (QELS) or sedimentation velocity. We have found that SEM or TEM analysis of solutions gives very detailed information about particle size distribution if a reproducible sample preparation method is found. We have found freeze-drying a sample spread onto the substrate from dilute solution to be effective. Static light scattering is the method of choice to determine the weight average molecular weight of these micelles, which runs into the millions (i.e. aggregation numbers in the hundreds).

Surface Characterization: Characterization of the surfaces that have been modified by polymer micelle adsorption or covalent attachment usually involves SEM, TEM, AFM or measurements of contact angle.

RECENT RESULTS

In the previous section the following progression was described in broad overview:

water soluble polymer → adsorbed polymer → self-assembled amphiphilic polymer

In this section we describe our most recent work in more detail.

POLYMER ADSORPTION ONTO LATEX PARTICLES

The polymers used in these studies contain a small mole fraction of hydrophobic chromophore covalently attached to a water soluble polymer:

CH$_3$ COOH

$\left(C-CH_2 \right) \left(C-CH_2 \right)$

(CH$_2$)$_2$-O-C=O   .01   CH$_3$ 0.99

H

PMA polymers

(CH$_2$)-(O-CH$_2$CH$_2$-)$_n$O-CH$_3$

H

PEO polymers

The polystyrene latex particles are commercially available ("microspheres" from Polysciences) with very monodisperse hydrodynamic diameters (60 nm, polydispersity based on quasi-elastic light scattering (QELS) is less than 0.05). Because of their relatively small size the area to volume ratio is favorable and the turbidity for a given concentration of latex is not too high. These considerations argue against the more commonly available commercial latexes with diameters > 150 nm.

Adsorption of the polymer onto the latex is easily followed by QELS such that the apparent adsorption isotherm can be obtained. We have also found fluorescence quenching to be very useful because the efficiency of fluorescence quenching by simple ions (e.g. Tℓ^+, I$^-$) or SPV changes at different points in the adsorption isotherm in ways that elucidate exposure of the chromophore to the aqueous phase.

Studies of the photoredox processes follows the methodology described earlier. Some of these systems have shown the highest singlet state ϕ_{cs} we have observed, especially at high pH (see Table 1). These studies do demonstrate that polymer surfaces can be used to provide the required hydrophobic protection for singlet state change separation. There was also the unexpected observation that PEO$-^1$D* in homogenous solution can yield ion-pairs with an efficiency (ca. 30%) that is as high or higher than when adsorbed onto latexes. Thus PEO-based polymers seem to be an excellent system for homogenous solution photoredox process.

AMPHIPHILIC BLOCK POLYMERS

There are various methods that place chromphores on strategic positions in a diblock polymer prepared by anionic polymerization: 1) the chromophore can be incorporated into the initiator, 2) the chromphore can be polymerized as one of the blocks, 3) the chromophore can be inserted between blocks as a analog of 1,1 diphenylethylene, 4) the chromophore can be used as a terminator. Since we are primarily interested in placing the chromophore near the hydrophobe-water interface methods 2) and 3) have been employed to produce species **2** and **3**.

CH$_3$

R-C-(CH$_2$CH) -CH$_2$ C $\left(CH_2-C \right)_m$ H

CH$_3$ C$_6$H$_5$ CH$_3$ COOH

2

CH$_3$

Cumyl $\left(CH_2-CH \right)_p \left(CH_2-CH \right)_n \left(CH_2-C \right)_m$ H

COOH

3

Polymers analogous to **2** have also been prepared with 2-vinyl-naphthalene, which is not suitable for photoredox studies, but is more tractable in synthesis. All these polymers self-assemble into micelles, following dialysis from dioxane:water (typically 80 volume % dioxane) to 100% water (typically with buffer or LiCl as a source of ionic strength). The exact details of micelle formation depends on polymer composition and molecular weight. These micelles have been proved to be very rugged, lasting years as a stable solution or being able to be redispersed into water after freeze-drying. As was discussed earlier (Table 1), for micelles of species **2** and **3** the preliminary estimates of the change separation efficiencies are relatively low (ca. 5%) but a very long lifetime for the ion-pair was observed. In fact, steady state irradiation at 400 nm for a period of hours produced a measurable $SPV^{-\bullet}$ spectrum on a standard UV-Vis spectrophotometer, which implies a lifetime of minutes or longer. These spectra also reveal that the anthracene cation radical is not present (this was also found to be the case on the μs time scales, but there is no irreversible loss of the anthracene. This implies that some adventitious sacrificial reagent is present to react with the cation radical, i.e.

$$A^{+\bullet} + R \rightarrow A + R^{+\bullet} \text{ (decomposes)} \qquad (10)$$

We have previously observed a similar phenomenon with pyrene cation radicals $(Py^{+\bullet})$ and postulated that a photo Kolbe reaction between the $Py^{+\bullet}$ and nearby -COOH groups.[5] Perhaps the same mechanism is operative here, which would explain the exceptional lifetime of the $SPV^{-\bullet}$ species.

There are many other features of these novel photoactive polymer micelles that are under investigation.[3] For example, the anthracene triplet state is not quenched by SPV under conditions where facile singlet state quenching is observed. This is unprecedented in our experience and must imply that the local environment has made the driving force (which is higher for the singlet than the triplet by ca. 1 V) much more important than in homogeneous solution. We believe this is consistent with relatively long-range electron transfer in a rigid or very viscous environment.

SUMMARY

The modification of polymers to influence photochemical processes has been explored by relatively few workers but is full of promise.[8] One of the strengths of this area, as in macromolecular science in general, is the potential for synthesizing new polymers that contain arrays of chromophores in an environment that can be manipulated by the remaining polymer structure. While it is doubtful that the exquisite structures found in photosynthetic photon transduction systems can ever be duplicated in "man-made" polymers, it is likely that robust materials can be developed that lend themselves to device fabrication. It is also reasonable to think that there may be spin-offs from this work that lie outside the area of photochemical energy conversion, such as optical information storage.

REFERENCES

1 All the work described in this article has been supported by the United States Department of Energy, Chemistry Division (Basic Sciences)

2 The following are general references to polymer micelles: (a) "Fluorescence Polarization Study of Polymer Micelles", K Prochàzka, D J Kiserow, S E Webber (1995) Acta Polymerica 46: 277-290 ; (b) Tuzar Z, Kratochvíl P (1976) Adv Colloid Interface Sci 6:201; (c) Tuzar Z , Kratochvíl P (1993) "Micelles of Block and Graft Copolymers", in Surface and Colloid Science, E Matijevic, ed, 15, Plenum Press, p 1 - 83; (d) Riess G , Heuter G , Bahadur P (1985) in *Encyclopedia of Polymer Science and Engineering*, Vol 2, (Mark H ; Bikales N M ; Overberger C G ; Menges G , eds) 2nd Ed, p 324 Wiley -Interscience New York; (e) Quirk, R P ; Perry, S ; Medicuti, F ; Mattice, W L (1988) Macromolecules 21:2294; (f) Quirk, R P ; Zhu, L (1989) Makromol Chem 190:487; (g) Quirk, R P Schock, L E (1991) Macromolecules, L E Macromolecules 24: 1237; (h) Caldérava, F ; Mruska, Z ; Hurtrez, G ; Lerch, J -P ; Nugay, T ; Riess, G (1994) Macromolecules 27:1210; (i) Ni, S ; Juhué, D ; Moselhy, J ; Wang, Y ; Winnik, M A (1992) Macromolecules 25:496; (j) Chan, L ; Winnik, M A ; Al-Takrity, E T A ; Jenkins, A D ; Welton, D R M (1987) Makromol Chem 108: 2621; (k) Hruska, Z ; Vuillemin, B ; Riess, G ; Katz, A ; Winnik, M A (1992) Makromol Chem 193:1987

3 Manuscript in preparation and dissertation, Andrew R. Eckert (1996), U. of Texas

4 Lakowicz, J R (1983) "Principles of Fluorescence Spectroscopy", Plenum Press, New York,, Chapter 9, especially p 279-282

5 Hsiao, J-S ; Webber (1992) J Phys Chem, 96:2892

6 Cayias, J L; Schechter, R S ; Wade, W H (1975) ACS Symposium Series, No 8, *Adsorption at Interfaces*, 234-247

7 See, for example, P C Hiemenz (1986), *Principles of Colloid and Surface Chemistry*, Marcel Dekker, Inc, p 385-398

8 This area has been reviewed by Y. Morishima in the article (1992) "Photoinduced Electron Transfer in Amphiphilic Polyelectrolyte Systems" Adv. Polym. Sci. 104:52

Chemical Functions of NH---S Hydrogen Bonds in Model Complexes of Iron-Sulfur Metalloproteins

Norikazu Ueyama, Taka-aki Okamura, Akira Nakamura

Department of Macromolecular Science, Faculty of Science,
Osaka University, Osaka 560 Japan

Abstract: Rubredoxin peptide model complex, [Fe^{II}(Z-**cys**-Pro-Leu-**cys**-Gly-Val-OMe)$_2$]$^{2-}$ containing an Fe ion surrounded by four cysteine thiolates of two invariant Cys-X-Y-Cys (X, Y = amino acid residues) fragments, exhibits the presence of NH---S hydrogen bonds between Cys sulfur and amide NH which is supported in a low dielectric constant solvent (e.g. 1,2-dimethoxyethane). The hydrogen bonding effectively contributes to the positive shift of redox potential. [Fe^{II}(Z-**cys**-Pro-Leu-**cys**-Gly-NHC$_6$H$_4$-*p*-X)$_2$]$^{2-}$ and [Fe^{II}(Z-**cys**-Pro-Leu-**cys**-Gly-Phe-OMe)$_2$]$^{2-}$ exhibit the presence of cooperative interaction between aromatic ring and NH---S hydrogen bonded sulfur atom. Simple model complexes contains singly and doubly NH---S hydrogen bonded thiolate ligands, e.g. 2-acylaminobenzenethiolate, 2,6-diacylaminobenzene-thiolate, capable of adopting a preferable structure for the intramolecular hydrogen bond. The crystallographic analysis of [Fe^{II}(S-2-*t*-BuCONHC$_6$H$_4$)$_4$]$^{2-}$ indicates that the hydrogen bond shortens the Fe-S bond presumably by influencing the antibonding SOMOs. Furthermore, the redox potentials of artificial model complexes, [Fe^{II}(Z-**cys**-Pro-Leu-**cys**-Gly-NHC$_6$H$_4$-*p*-X)$_2$]$^{2-}$, having a *p*-substituted (X = OMe, H, F, CN) anilide group obey to the Hammett rule with the σ_p values. It indicates that the electronic effect of *p*-substituents extends to Fe(II) ion through the benzene ring and the NH---S hydrogen bond. Thus, NH---S hydrogen bond is one of the candidates for electron pathway in the active site of electron transfer iron-sulfur proteins.

Introduction

Some of electron-transfer metalloproteins contain an iron or copper ion surrounded by many Cys thiolate ligands in the active center. The presence of NH---S hydrogen bond has been proposed by the X-ray crystallographic analysis for these metalloproteins. For example, rubredoxins contain two double NH---S hydrogen bonds and two single ones [1]. Similarly, the active sites of plant-type ferredoxins [2], bacterial ferredoxins [3] and blue-copper proteins [4] have been proposed to have both types of the hydrogen bonds as shown in Fig. 1.

In general, one metal ion at the active site of metalloproteins is covered by 8~10 amino acid residues and this part is furthermore surrounded by 60~100 amino acid residues to dissolve in aqueous solution. A recent graphic analysis of these metalloproteins has revealed that the active site is surrounded by hydrophobic amino acid side chains with a highly hydrophobic contrast [5]. These proteins were found to have a unique layer structure consisting of three layers, an

M. Kamachi · A. Nakamura (Eds)
New Macromolecular Architecture and Functions
Proceedings of the OUMS '95 Toyonaka, Osaka, Japan, 2-5 June, 1995
© Springer-Verlag Berlin Heidelberg 1996

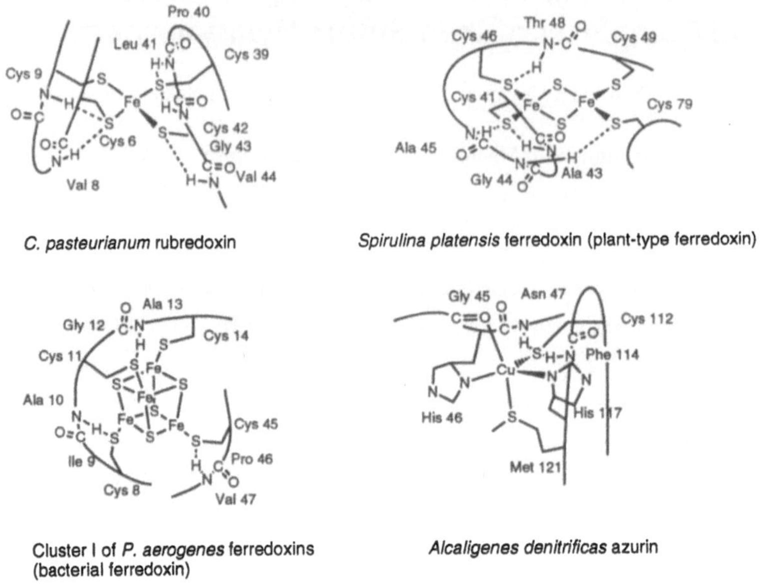

148

C. pasteurianum rubredoxin

Spirulina platensis ferredoxin (plant-type ferredoxin)

Cluster I of P. aerogenes ferredoxins
(bacterial ferredoxin)

Alcaligenes denitrificas azurin

Fig. 1. The proposed NH---S hydrogen bonds in the active sites of rubredoxin, plant-type [2Fe-2S] ferredoxin, bacterial [4Fe-4S] ferredoxin and azurin.

electrostatic layer including the metal-binding site, a hydrophobic layer with nonpolar amino acid residues directed to the metal ion core and a hydrophilic layer with polar amino acid residues on the surface. The hydrophobic environments produce a domain of low dielectric sphere and enhance the complexation between cationic metal ion and anionic ligands. The hydrophilic layer assists the protein in dissolving in water. By the above-mentioned heterogeneous structure, many difficulties arise in the chemical simulation of these metalloproteins.

To solve these problems, we have synthesized oligopeptide model complexes without the surface hydrophilic layer. The solution of the complexes in nonpolar organic solvent can

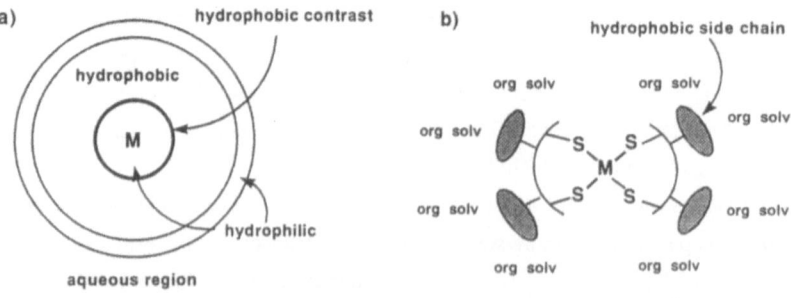

Fig. 2. Schematic drawings of the structure for a) metalloproteins and b) synthetic oligopeptide model complexes.

simulate its chemical environments around the active center of metalloproteins as shown in Fig.2. A series of peptide model complexes containing each invariant sequence at the [1Fe] core in rubredoxin [6-8], the [2Fe-2S] core in plant-type ferredoxin [9] or the [4Fe-4S] core in ferredoxin [10] have been synthesized. The peptide model complexes are found to have a preferable conformation to readily form the NH---S hydrogen bond.

Positive Shift of the Redox Potential by NH---S Hydrogen Bond in Rubredoxin Peptide Model Complexes

One of the simplest electron transfer iron-sulfur proteins, rubredoxin, has an Fe ion surrounded by four cysteine thiolates of two invariant Cys-X-Y-Cys (X, Y = amino acid residues) fragments [11, 12]. Various rubredoxin model complexes having a chelating tetrapeptide ligand, $[Fe^{II}(Z\text{-}cys\text{-}X\text{-}Y\text{-}cys\text{-}OMe)_2]^{2-}$ (X-Y = Ala-Ala, Pro-Leu, Thr-Val) were synthesized to construct spectral and electrochemical models having characteristic properties of native rubredoxin [6]. Larger peptide complexes were also synthesized as models containing a series of Cys-X-Y-Cys-Gly-A hexapeptide fragments existing within 3 Å of the Fe ion [7, 8, 13, 14].

Figure 3 shows the ^{2}H-NMR spectra of a deuterated amide N^2H-peptide complex, $[Fe^{II}(Z\text{-}cys\text{-}Pro\text{-}Leu\text{-}cys\text{-}Gly\text{-}Val\text{-}OMe)_2]^{2-}$, in acetonitrile at room temperature. The spectra indicate the presence of two N^2H--S hydrogen bonds in the chelating Cys-Pro-Leu-Cys moiety and one such hydrogen bond by the Cys-Gly-Val-OMe fragment [8]. The contact shifted amide N^2H signals are observed at 40 ppm, 20 ppm and -20 ppm. These shifts mainly due to a Fermi contact clearly indicate the presence of the NH---S hydrogen bond of partially covalent nature.

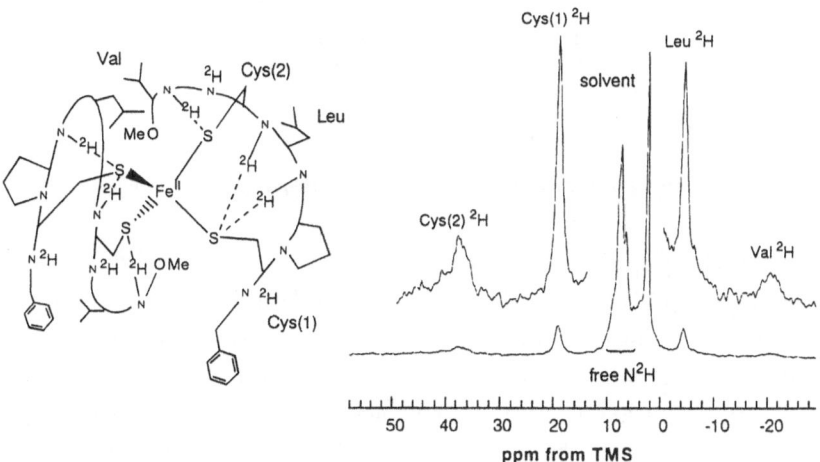

Fig. 3. ^{2}H NMR spectrum of N^2H-$[Fe^{II}(Z\text{-}cys\text{-}Pro\text{-}Leu\text{-}cys\text{-}Gly\text{-}Val\text{-}OMe)_2]^{2-}$ in acetonitrile at room temperature.

The Cys-Gly-Val-OMe unit provides a turn structure to give the specific NH---S hydrogen bond in a low-dielectric-constant solvent. In stead of the Val residue, p-substituted anilide was introduced in the tripeptide fragment like Cys-Gly-NHC$_6$H$_4$-p-X (X = OMe, H, F, CN) to investigate the distal electronic effect through the NH---S hydrogen bond [7].

Table 1. 2-/3- Redox Potentials of Fe(II) Peptide Complexes in 1,2- dimethoxyethane (DME) at Room Temperature [7].

Complexes	σ_p	Redox potential, V vs SCE
[Fe^{II}(Z-**cys**-Pro-Leu-**cys**-OMe)$_2$]$^{2-}$		-0.59
[Fe^{II}(Z-**cys**-Pro-Leu-**cys**-Gly-Val-OMe)$_2$]$^{2-}$		-0.35
[Fe^{II}(Z-**cys**-Pro-Leu-**cys**-Gly-NHC$_6$H$_4$-p-OMe)$_2$]$^{2-}$	-0.27	-0.33
[Fe^{II}(Z-**cys**-Pro-Leu-**cys**-Gly-NHC$_6$H$_5$)$_2$]$^{2-}$	0	-0.33
[Fe^{II}(Z-**cys**-Pro-Leu-**cys**-Gly-NHC$_6$H$_4$-p-F)$_2$]$^{2-}$	0.06	-0.31
[Fe^{II}(Z-**cys**-Pro-Leu-**cys**-Gly-NHC$_6$H$_4$-p-CN)$_2$]$^{2-}$	0.66	-0.24

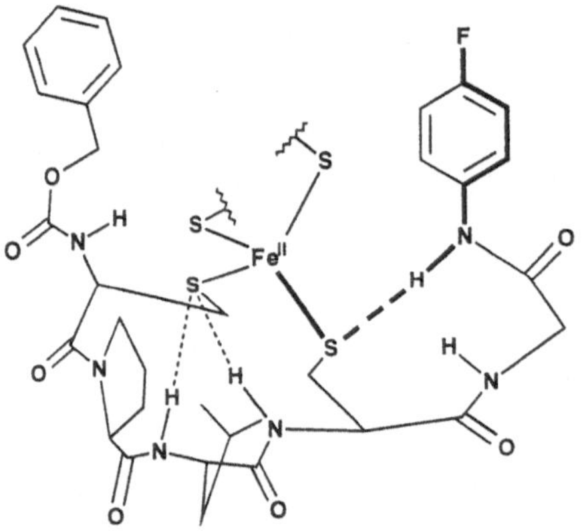

Fig. 4. Schematic structure of [Fe^{II}(Z-**cys**-Pro-Leu-**cys**-Gly-NHC$_6$H$_4$-p-F)$_2$]$^{2-}$ in acetonitrile.

The electrochemical properties of these peptide model complexes were studied using cyclic voltammograms as listed in Table 1. The positive-shifted redox potential at -0.35 V vs SCE for [Fe^{II}(Z-**cys**-Pro-Leu-**cys**-Gly-Val-OMe)$_2$]$^{2-}$ in DME is close to that (-0.06 V vs NHE and -0.31 V vs SCE) reported for *Clostridium pasteurianum* rubredoxin [11]. The complexes having various kinds of p-substituted (OMe, H, F, CN) anilide derivatives exhibit the redox potentials obeying to a Hammett rule with the σ_p values. The results indicates that the electronic effect of p-substituents extends to Fe(II) ion through the benzene ring and the NH---S hydrogen bond (Fig. 4).

The ^{2}H NMR spectra of [Fe^{II}(Z-**cys**-Pro-Leu-**cys**-Gly-NHC$_6$H$_4$-p-F)$_2$]$^{2-}$ in acetonitrile show contact-shift signals at 40 ppm, 20 ppm and -10 ppm at -30°C similar to those of the above hexapeptide model complex. The ^{19}F NMR spectra of [Fe^{II}(Z-**cys**-Pro-Leu-**cys**-Gly-NHC$_6$H$_4$-p-F)$_2$]$^{2-}$ show signals at -122.6 ppm and -122.9 ppm (free liagnd ^{19}F signal at 199.9 ppm) due to the presence of two isomers as described later [13, 15]. The corresponding m-fluoro derivative, [Fe^{II}(Z-**cys**-Pro-Leu-**cys**-Gly-NHC$_6$H$_4$-m-F)$_2$]$^{2-}$, exhibits the corresponding ^{19}F signals at -103.5 ppm -108.3 ppm (free ligand ^{19}F signal at -114.0 ppm). Each signal for the p-^{19}F and m-^{19}F in the two complexes shows the opposite shift by paramagnetic Fe(II). Thus, the unpaired electron delocalizes from Fe(II) ion through the NH---S hydrogen bond and then the antibonding molecular orbital on the benzene ring and reaches the ^{19}F atom.

Two Isomers by Coordination of Cys-X-Y-Cys

Tetrahedral Fe(II) complex of bidentate peptide ligand, e.g. Z-Cys-Pro-Leu-Cys-OMe, has two isomers (δ and λ) with the different orientations in coordination (Fig. 5a and 5b). Actually, a complex [Fe^{II}(Z-**cys**-Pro-Leu-**cys**-Gly-NHC$_6$H$_4$-m-F)$_2$]$^{2-}$ shows six Cys C$_\beta$H$_2$ proton signals at 271, 255, 238, 224, 209 and 156 ppm in acetonitrile-d_3 at 30 °C as well as in the case of [Fe^{II}(Z-Ala-**cys**-Pro-Leu-**cys**-Gly-NHC$_6$H$_4$-m-F)$_2$]$^{2-}$. The signals at ca. 271 and 255 ppm correspond to those of two isomers due to one of two enantiotopic protons of Cys C$_\beta$H$_A$H$_B$. These isomers were detected by the ^{19}F NMR spectra as described above. On the contrary, [Fe^{II}(Z-**cys**-Pro-Leu-**cys**-Gly-Val-OMe)$_2$]$^{2-}$ exhibit four separate Cys C$_\beta$H$_2$ signals at 257, 215, 200 and 147 ppm. To assign these signals, we synthesized two tetradentate peptide ligands, cis-1,2-cyclohexylene(CO-Cys-Pro-Leu-Cys-NHC$_6$H$_4$-m-F)$_2$ and cis-1,2-cyclohexylene(CO-Ala-Cys-Pro-Leu-Cys-NHC$_6$H$_4$-m-F)$_2$ by considering two different sizes in the spacers, cis-1,2-cyclohexylene(CO-)$_2$ and cis-1,2-cyclohexylene(CO-Ala-)$_2$ [15, 16]. The ^{1}H and ^{2}H NMR spectra of both complexes show the existence of only one isomer (δ form).

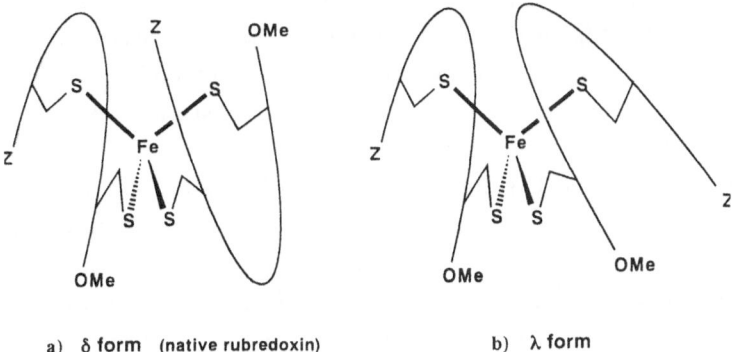

a) δ **form** (native rubredoxin)   b) λ **form**

Fig. 5.   Schematic illustrations of a) δ and b) λ forms as the two isomers in [Fe^{II}(Z-**cys**-Pro-Leu-**cys**-Gly-Val-OMe)$_2$]$^{2-}$.

Table 2. Isomer Contents of δ and λ forms in Various Bidentate Peptide Model Complexes.

Complexes	Ratios $(\delta : \lambda)$
$[Fe^{II}(Z\text{-}\mathbf{cys}\text{-}Pro\text{-}Leu\text{-}\mathbf{cys}\text{-}OMe)_2]^{2-}$	1 : 1
$[Fe^{II}(Z\text{-}\mathbf{cys}\text{-}Pro\text{-}Val\text{-}\mathbf{cys}\text{-}OMe)_2]^{2-}$	7 : 3
$[Fe^{II}(Z\text{-}\mathbf{cys}\text{-}Pro\text{-}Leu\text{-}\mathbf{cys}\text{-}Gly\text{-}Val\text{-}OMe)_2]^{2-}$	1 : 0
$[Fe^{II}(Z\text{-}\mathbf{cys}\text{-}Pro\text{-}Leu\text{-}\mathbf{cys}\text{-}Gly\text{-}NHC_6H_4\text{-}m\text{-}F)_2]^{2-}$	1 : 1
$[Fe^{II}(Z\text{-}Ala\text{-}\mathbf{cys}\text{-}Pro\text{-}Leu\text{-}\mathbf{cys}\text{-}Gly\text{-}NHC_6H_5)_2]^{2-}$	6 : 4
$[Fe^{II}\{cis\text{-}1,2\text{-}cyclohexylene(CO\text{-}Ala\text{-}\mathbf{cys}\text{-}Pro\text{-}Leu\text{-}\mathbf{cys}\text{-}Gly\text{-}NHC_6H_4\text{-}m\text{-}F)_2\}]^{2-}$	1 : 0
$[Fe^{II}\{cis\text{-}1,2\text{-}cyclohexylene(CO\text{-}Ala\text{-}\mathbf{cys}\text{-}Pro\text{-}Leu\text{-}\mathbf{cys}\text{-}Gly\text{-}NHC_6H_4\text{-}m\text{-}F)_2\}]^{2-}$	1 : 0

For example, $[Fe^{II}\{cis\text{-}1,2\text{-}cyclohexylene(CO\text{-}Ala\text{-}\mathbf{cys}\text{-}Pro\text{-}Leu\text{-}\mathbf{cys}\text{-}Gly\text{-}NHC_6H_4\text{-}m\text{-}F)_2\}]^{2-}$ exhibits four signals at 254, 222, 208 and 160 ppm. These signals also correspond to the four Cys $C_\beta H_2$ signals repored for native *Desulfovibrio gigas* rubredoxin [17]. Table 2 lists the ratios of the δ and λ isomers analyzed by the ^1H and ^2H NMR signals.

The presence of only one δ isomer in $[Fe^{II}(Z\text{-}\mathbf{cys}\text{-}Pro\text{-}Leu\text{-}\mathbf{cys}\text{-}Gly\text{-}Val\text{-}OMe)_2]^{2-}$ is also established by the ^2H NMR analysis. The δ form is more stable than the λ form based on the difference of 16 kcal/mol in energy-minimized total energy between two isomers calculated using a Dreiding force field. The steric congestion between two Gly-Val-OMe groups in $[Fe^{II}(Z\text{-}\mathbf{cys}\text{-}Pro\text{-}Leu\text{-}\mathbf{cys}\text{-}Gly\text{-}Val\text{-}OMe)_2]^{2-}$ permits the existence of the δ form only. Thus, it is interesting that steric restriction for the formation of only one δ isomer is induced by oligopeptide fragments consisting of L-amino acid residues (tetra or hexapeptide) around the binding site of metal ion in rubredoxins.

In the case of plant-type ferredoxin model complexes containing an $(Fe_2S_2)^{2+}$ core, two coordination isomers are also formed. For example, we synthesized a novel oligopeptide model complex, $(NEt_4)_2[Fe_2S_2(20\text{-}pep)]$ (20-pep = Ac-Pro-Tyr-Ser-Cys-Arg-Ala-Gly-Ala-Cys-Ser-Thr-Cys-Ala-Gly-Pro-Leu-Leu-Thr-Cys-Val-NH_2) having an invariant amino acid residues around the active center of *Spirulina platensis* ferredoxins. The complex exhibited two redox couples at -0.64 and -0.96 V vs SCE in DMF due to the presence of two isomers.. The former of them has a similar redox potential in $(Fe^{II}Fe^{III}S_2)^+/(Fe^{III}_2S_2)^{2+}$ to that of native ferredoxins in aqueous solution [9].

A tridentate peptide fragment (Cys-Arg-Ala-Gly-Ala-Cys-Ser-Thr-Cys) chelates an $(Fe^{III}_2S_2)^{2+}$ core with two coordination modes as shown in Fig. 6. The positive shift of the redox potential for the former isomer (isomer A) has been proposed to be ascribed to the presence of single or double NH---S hydrogen bonds as shown by X-ray analysis of *S. platensis* ferredoxin [18]. On the contrary, the latter isomer (isomer B) has a negative redox potential without NH---S hydrogen bond due to the unfavourable conformation. In general, simple $(Fe_2S_2)^{2+}$ complexes having alkane- or arenethiolate ligands exhibit a negative redox potential in DMF, e.g. -1.49,

-1.09 and -1.31 V vs SCE for $[Fe_2S_2(S_2\text{-}o\text{-xyl})_2]^{2-}$, $[Fe_2S_2(SPh)_4]^{2-}$ and $[Fe_2S_2(S\text{-}t\text{-Bu})_4]^{2-}$, respectively [19].

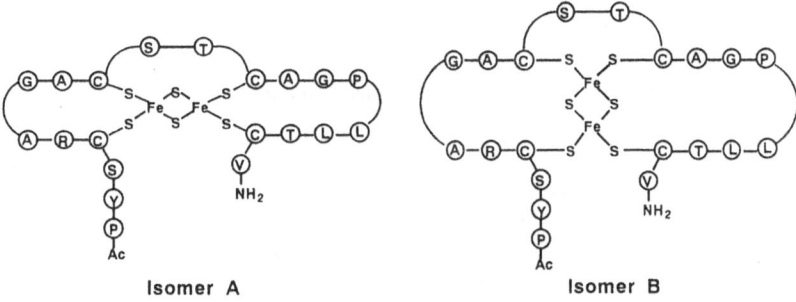

Isomer A   **Isomer B**

Fig. 6. Two possible chelations (isomer A and B) by the 20-pep ligand to an $(Fe_2S_2)^{2+}$ core.

Cooperative Stabilization of the Active Center between Aromatic Ring and NH---S Hydrogen Bond

Function of an aromatic ring around the active site of metalloproteins has been discussed by Pesko et al. [20]. At present, three chemical functions have been proposed for the metalloproteins. The first is the formation of hydrophobic environments to support electrostatic interactions by the aromatic rings around the metal center. The second is the function as an electron transfer mediator by taking a specific orientation among several aromatic and heterocyclic rings. The third is a direct $d\pi$–$p\pi$ interaction between metal ion and an aromatic ring [21].

New Cys-containing oligopeptide ligands containing an isolated aromatic ring at the C-terminus without π conjugation between the ring and the amide group were designed and the corresponding Fe(II) complexes were synthesized. Thus, $(NEt_4)_2[Fe^{II}(Z\text{-}\mathbf{cys}\text{-Pro-Leu-}\mathbf{cys}\text{-}$ Gly-$NHCH_2C_6H_4\text{-}p\text{-F})_2]$ and $(NEt_4)_2[Fe^{II}(Z\text{-}\mathbf{cys}\text{-Pro-Leu-}\mathbf{cys}\text{-Gly-}NHCH_2CH_2C_6H_4\text{-}p\text{-F})_2]$ were synthesized. Similarly, $[Fe^{II}(Z\text{-}\mathbf{cys}\text{-Pro-Leu-}\mathbf{cys}\text{-Gly-Phe-OMe})_2]^{2-}$ containing a peptide fragment with the phenyl group found in the active site of rubredoxin was also prepared. The ^2H-NMR results indicate the presence of the hydrogen bonds in the same manner as that of the above-mentioned hexapeptide model complexes.

Table 3 lists the absorption maxima and CD extrema of oligopeptide model complexes in acetonitrile. The introduction of the aromatic ring into the peptide complexes results in appearance of a new LMCT band at 330 ~ 332 nm. This band corresponds to the absorption maximum at 333 nm reported for native reduced rubredoxin. When the peptide ligand has an aromatic ring in a suitable place, a remarkably large $\Delta\varepsilon$ value of the CD extrema for these peptide complexes are observed. The complex, $[Fe^{II}(Z\text{-}\mathbf{cys}\text{-Pro-Leu-}\mathbf{cys}\text{-Gly-Phe-OMe})_2]^{2-}$, exhibits a large $\Delta\varepsilon$ (-27) similar to the value ($\Delta\varepsilon$ = -36) of native reduced rubredoxin. Most of synthetic peptide model complexes have shown only small $\Delta\varepsilon$ extrema that have been considered to be ascribed to the non-rigidity of the peptide conformation around the metal center. However, the aromatic ring cooperates with the NH---S hydrogen bond to give a rigid structure even in the short peptide ligand. The formation of NH---S hydrogen bonds are also supported by the contact-shifted N^2H NMR signals at 40.5, 35.7, 18.8, -2.4 and -4.6 ppm in acetonitrile at 30°C. The temperature dependence of chemical shifts of these signals is subject to a Curie-Weiss law. This behavior indicates that the distant amide NH forms a NH---S hydrogen bond with Cys thiolate. The adjacent aromatic ring on the Phe residue is thought to interact directly with the hydrogen bonded sulfur atom.

Table 3. UV-visible and CD Spectral Data for Fe(II)-Peptide Complexes in Acetonitrile and Rubredoxin in Aqueous Solution.

Complexes	UV-vis maxima[a]	CD extrema[b]	
$[Fe^{II}(Z$-**cys**-Gly-Val-OMe)$_2]^{2-}$	316(5300)	320(-3.5)	341(1.2)
$[Fe^{II}(Z$-**cys**-Pro-Val-**cys**-OMe)$_2]^{2-}$	314(5900)	320(-6.3)	340(3.3)
$[Fe^{II}(Z$-**cys**-Pro-Leu-**cys**-Gly-Val-OMe)$_2]^{2-}$	312(4500)	309(-5.3)	338(2.8)
$[Fe^{II}(Z$-**cys**-Pro-Leu-**cys**-Gly-NHC$_6$H$_4$-p-F)$_2]^{2-}$	312(5900) 332(4800)	308(-10) 322(-9.6)	339(6.9)
$[Fe^{II}(Z$-**cys**-Pro-Leu-**cys**-Gly-NHCH$_2$C$_6$H$_4$-p-F)$_2]^{2-}$	312(6780) 331(6030)	315(-12)	335(6.0)
$[Fe^{II}(Z$-**cys**-Pro-Leu-**cys**-Gly-NHCH$_2$CH$_2$C$_6$H$_4$-p-F)$_2]^{2-}$	312(7400) 330(6200)	316(-19)	335(9.0)
$[Fe^{II}(Z$-**cys**-Pro-Leu-**cys**-Gly-Phe-OMe)$_2]^{2-}$	312(8240) 332(6750)	316(-27)	334(16)
Reduced rubredoxin[c]	312(10900)333(6000)	314(-36)	334(18)

[a] In nm (ε, M^{-1}cm^{-1}). [b] In nm ($\Delta\varepsilon$, M^{-1}cm^{-1}). [c] Eaton and Lovenberg [12].

(NEt$_4$)$_2$[Fe^{II}(Z-**cys**-Pro-Leu-**cys**-Gly-NHCH$_2$C$_6$H$_4$-p-F)$_2$] and (NEt$_4$)$_2$[Fe^{II}(Z-**cys**-Pro-Leu-**cys**-Gly-NHCH$_2$CH$_2$C$_6$H$_4$-p-F)$_2$] exhibit slightly contact shifted ^{19}F signals due to a dipolar contact which does not obey the Curie-Weiss law [14]. The sulfur atom probably prefers a tetrahedral structure having two types of interactions with the amide NH group and the aromatic ring. In native metalloproteins, aromatic ring at Tyr or Phe residue cooperating with the NH---S hydrogen bond covers over the sulfur atom to presumably prevent hydrolysis by water and oxidation by air.

Intramolecular Single and Double NH---S Hydrogen Bonds in Simple Model Complexes

Novel metal thiolate complexes with intramolecular singly or doubly hydrogen bonded ligands were designed to investigate the more detailed chemical functions of NH---S hydrogen bond (Fig. 7). In metalloproteins, Cys thiolate anion is hydrogen-bonded with the amide NH groups of the adjacent amino acid residues which are supported by a helix or β-sheet conformation. The designed ligands contain rigid amide NH groups to form intramolecular NH---S hydrogen bonds. Mononuclear Fe(II) and Co(II) complexes having the above thiolate ligands, e.g. (NEt$_4$)$_2$[Fe^{II}(S-2-t-BuCONHC$_6$H$_4$)$_4$] and (NEt$_4$)$_2$[Co^{II}(S-2-t-BuCONHC$_6$H$_4$)$_4$], were

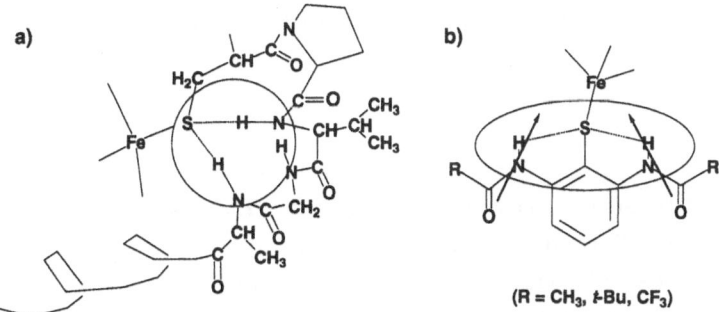

Fig. 7. a) NH---S hydrogen bonded Cys thiolate adjacent to a helix in metalloproteins. b) Designed arenethiolate having an intramolecular double NH---S hydrogen bond. The arrow refers to amide dipole.

synthesized as simple peptide-mimetic models of rubredoxin [22], different from the intermolecular NH---S hydrogen bonds between thiolate and amide cation in the solid state reported for $(Me_3NCH_2CONH_2)_2[Co^{II}(SPh)_4]$ [23]. The crystal structures of $(NEt_4)_2[Fe^{II}(S-2-t-BuCONHC_6H_4)_4]$ and $(NEt_4)_2[Co^{II}(S-2-t-BuCONHC_6H_4)_4]$ show a tetrahedral geometry for the MS_4 cores with the direction of each NH group to the sulfur atoms as shown in Fig. 8. Each of the bond

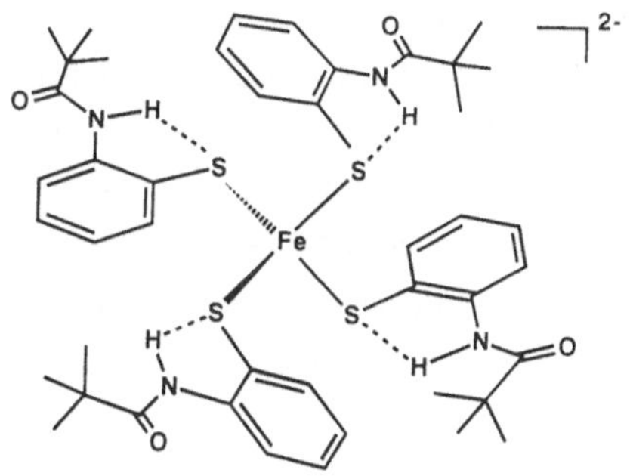

Fig. 8. Structure of $[Fe^{II}(S-2-t-BuCONHC_6H_4)_4]^{2-}$.

distances of Fe-S (2.329 Å) and Co-S (2.296 Å) in these complexes is shorter than that of the corresponding $[Fe^{II}(SPh)_4]^{2-}$ (2.353 Å) [24] and $[Co^{II}(SPh)_4]^{2-}$ (2.328 Å) [25] complexes. The observed shortening at the metal-sulfur bonds can be explained as follows. The singly-occupied orbitals (SOMO) in these complexes are of mostly d-character and are antibonding combination of metal d and sulfur p orbitals. As the NH---S hydrogen bonding occurs to the sulfur π-orbital, the electron-density of the SOMO will be decreased depending on the strength of the hydrogen bond.

The IR spectrum in the solid state of $(NEt_4)_2[Co^{II}(S-2-t-BuCONHC_6H_4)_4]$, which has a similsr electron configuration to the Fe complex, shows an amide NH band at 3281 cm^{-1} and a CO band at 1667 cm^{-1}. The corresponding disulfide, $(S-2-t-BuCONHC_6H_4)_2$, in the solid state gives a free NH band at 3389 cm^{-1} and an intermolecularly NH---O=C hydrogen bonded amide NH band at 3251 cm^{-1} separately. On the other hand, a free amide CO band and an intermolecularly hydrogen bonded CO were observed at 1679 cm^{-1} and 1644 cm^{-1}, respectively. No observation of these bands in the Co(II) complex indicates the absence of both the free NH and the intermolecularly NH---O=C hydrogen bonded amide NH bands. The band at 3281 cm^{-1} is assignable to the amide NH stretching in the NH---S hydrogen bond.

$(NEt_4)_2[Fe^{II}(S-2-t-BuCONHC_6H_4)_4]$ exhibits a redox potential at -0.29 V vs. SCE in acetonitrile. The positive shift by the NH---S hydrogen bonding is 0.24 V as compared with the value, -0.53 V vs SCE, for $(NEt_4)_2[Fe^{II}(SPh)_4]$.

Similarly, the crystal structure of Co(II) thiolate complex, $[Co^{II}\{S-2,6-(CF_3CONH)_2C_6H_3\}_4]^{2-}$, having intramolecular double NH---S hydrogen bonds indicates that all of the amide NH groups are directed to the sulfur atoms. The presence of NH---S hydrogen bonds was also confirmed by the IR spectra in the solid state. When compared to that of $[Co^{II}(SPh)_4]^{2-}$, the Fe-S bond distance of the complex having double NH---S hydrogen bonds is shortened by approximately 0.03 Å. The NH---S hydrogen bonding is considered to decrease the electron density of Co-S $d\pi$-$p\pi$ SOMO as described above.

The crystal structures of molybdenum(IV) and copper(I) complexes, $(PPh_4)_2[Mo^{IV}(S-2-CH_3CONHC_6H_4)_4]$ [26] and $(NEt_4)_2[Cu^I(S-2-t-BuCONHC_6H_4)_3]$ [27], indicate the shortening of the M-S (M = Mo(IV), Cu(I)) bond distances. The large IR shift of the amide NH stretching in $(NEt_4)_2[Cu^I(S-2-t-BuCONHC_6H_4)_3]$ indicates that the NH---S hydrogen bond is stronger than that of a NH---O=C hydrogen bond. Thus, in the case of biologically significant metal-thiolate proteins, the NH---S hydrogen bond affects the antibonding HOMO of metal-sulfur core and results in strengthening the M-S bond. This effect contributes to the large metal-binding constant in metalloproteins.

Conclusions

Biologically and chemically important functions of NH---S hydrogen bond in iron-sulfur metalloproteins have been investigated using oligopeptide and simple model complexes. The hydrogen bonding contributes to the positive shift of redox potential, to the cooperative interaction between the NH---S hydrogen bonded sulfur and aromatic ring and to the strengthening of Fe-S bond. Thus, the amide NH interacts with sulfur $p\pi$ and affects the Fe-S bond. We will still continue our effort to find other chemical functions caused by the NH---S hydrogen bond.

The electronic effect of p-substituents extends to Fe(II) ion through the benzene ring and the NH---S hydrogen bond. Thus, NH---S hydrogen bond is one of the candidates for electron pathway in the active site of electron transfer iron-sulfur proteins. Furthermore, it is likely that the invariant Cys-X-Y-Cys chelating ligand can regulate the electrochemical properties with its flexible structure when the electron transfer Cys-containing metalloproteins associate with other electron-transfer partner proteins.

References:

1. Watenpaugh K D, Sieker L C, Jensen L H (1979) J Mol Biol 131:509.
2. Tsukihara T, Fukuyama K, Nakamura M, Katsube Y, Kanaka N, Kakudo M, Hase T, Wada K, Matsubara H (1981) J Biochem 90:1763.

3. Adman E, Watenpaugh K D, Jensen L H (1975) Proc Natl Acad Sci USA 72:4854.

4. Baker E N (1988) J Mol Biol 203:1071.

5. Yamashita M M, Wesson L, Eisenman G, Eisenberg D (1990) Proc Natl Acad Sci USA 87:5648.

6. Ueyama N, Nakata M, Fuji M, Terakawa T, Nakamura A (1985) Inorg Chem 24:2190.

7. Sun W, Ueyama N, Nakamura A (1991) Inorg Chem 30:4027.

8. Ueyama N, Sun W Y, Nakamura A (1992) Inorg Chem 31:4053.

9. Ueyama N, Ueno S, Nakamura A, Wada K, Matsubara H, Kumagai S, Sakakibara S, Tsukihara T (1992) Biopolymers 32:1535.

10. Ohno R, Ueyama N, Nakamura A (1991) Inorg Chem 30:4887.

11. Lovenberg W, Sobel B E (1965) Proc Natl Acad Sci 54:193.

12. Eaton W A, Lovenberg W (ed. Lovenberg W) (1873) in Iron-Sulfur Proteins Academic Press, New York

13. Sun W, Ueyama N, Nakamura A (1993) J Chem Soc Dalton Trans 1871.

14. Sun W, Ueyama N, Nakamura A (1993) Inorg Chem 32:1095.

15. Sun W, Ueyama N, Nakamura A (1993) Mag Res Chem 31:S34.

16. Sun W Y, Kajiwara A, Ueyama N, Nakamura A (1992) J Chem Soc Dalton Trans 3255.

17. Werth M T, D. M. Kurtz J, Moura I, LeGall J (1987) J Am Chem Soc 109:273.

18. Tsukihara T, Fukuyama K, Tahara H, Katsube Y, Matsuura Y, Tanaka N, Kakudo M, Wada K, Matsubara H (1978) J Biochem 84:1645.

19. Holm R H (1977) Acc Chem Res 10:427.

20. Burley S K, Petsko G A (1988) Adv Protein Chem 39:125.

21. Yamauchi O, Odani A (1985) J Am Chem Soc 107:5938.

22. Ueyama N, Okamura T, Nakamura A (1992) J Chem Soc Chem Commun 1019.

23. Walters M A, Dewan J C, Min C, Pinto S (1991) Inorg Chem 30:2656.

24. Coucouvanis D, Swenson D, Baenziger N C, Murphy C, Holah D G, Sfarnas N, Simopoulos A, Kostikas A (1981) J Am Chem Soc 103:3350.

25. Swenson D, Baenzinger N C, Coucouvanis D (1978) J Am Chem Soc 100:1932.

26. Ueyama N, Okamura T, Nakamura A (1992) J Am Chem Soc 114:8129.

27. Okamura T, Ueyama N, Nakamura A, Ainscough E W, Brodie A M, Waters J M (1993) J Chem Soc Chem Commun 1658.

Genetically Expressed Monodisperse α Helical Polypeptides

Jeffrey S. Bartlett[1], Richard J. Samulski[1], Yuhua Li[2], and Edward T. Samulski[2]

Gene Therapy Center[1], The University of North Carolina at Chapel Hill, Chapel Hill, NC 27599-7352, Department of Chemistry[2], The University of North Carolina at Chapel Hill, Chapel Hill, NC 27599-3290.

ABSTRACT

Four different chain lengths of monodisperse, α-helical polypeptides derived from the protein comprising the tail fibers of the bacteriophage Ur-λ were prepared by recombinant DNA techniques. The DNA which codes for the tail fiber segment has been replicated and inserted into *Escherichia Coli* bacteria. The desired proteins expressed by the bacteria were harvested and purified with a nickel affinity chromatography. The target proteins were characterized by gel electrophoresis and matrix-assisted laser desorption/ionization (MALDI) mass spectrometry. The secondary α-helical structure of the polypeptides was characterized by circular dichroism (CD) spectrophotometry.

INTRODUCTION

We introduced the concept of self-assembled, rodlike macromolecules on gold (1-3) by exploiting the gold-thiol covalent bonding used in conventional self assembly of alkane thiols (4). In that work we employed synthetic polypeptides (poly-γ-benzyl-L-glutamate) (PBLG) derivitized at the amino-terminus with the disulphide-containing reagent lipoic acid. The major motivations for using the rodlike macromolecule PBLG was that it is readily available, forms a robust α-helical secondary structure, and is a synthetic polypeptide with excellent solubility in organic solvents—due to the favorable thermodynamics of mixing of its flexible side chain with the solvent molecules. However, even when the assembly was carried out in an electric field that partially aligns PBLG's macrodipole moment (5), infrared dichroism analysis indicated that the PBLG helix-axis orientation distribution was rather broad (3,5)—the resulting self-assemblay of rodlike polymers is not uniformly aligned with the helix axes approximately normal to the gold substrate (as is the case for self assemblys of simple alkane thiols). One potential difficulty with the use of commercial PBLG in the self assembly is its very broad molecular weight distribution—a broad distribution of rod lengths—and in subsequent work in our laboratory we opted to focus on this difficiency of the synthetic polypeptide model system.

In general rodlike polymers with precise lengths, defined primary structures (specified, sequential chemical functionality), and robust secondary conformations, could be amenable to a variety of applications in polymer materials science and serve as model compounds for evaluating

M. Kamachi · A. Nakamura (Eds)
New Macromolecular Architecture and Functions
Proceedings of the OUMS '95 Toyonaka, Osaka, Japan, 2-5 June, 1995
© Springer-Verlag Berlin Heidelberg 1996

theory describing idealized rods. Unfortunately such desirable attributes, especially precision (monodisperse) polymers, can not be produced via synthetic chemical polymerization processes due to the statistical nature of such processes. In the case of polypeptides, however, recombinant DNA methods may be used to design and synthesize polypeptides with a specified primary structure exhibiting well-defined secondary and tertiary structures (6-8). It should be emphasized, however, that the expression of proteins from artificial genes may be problematic: Genes encoding repetitive polypeptides are of necessity comprised of repetitive DNA sequences that can be particularly unstable in bacterial hosts—repetitive messenger RNAs may adopt folded structures that are translated inefficiently or degraded rapidly (9). Previous attempts at expression of large repetitive genes have met with limited success (10).

Rather than having to face the compatibility questions associated with producing an artificial polypeptide in a bacterial expression system, herein we report an efficient procedure to express naturally occurring rodlike proteins. We have chosen to use the naturally occurring amino acid sequence comprising the ~35 nm long rodlike tail fiber of the bacteriophage Ur-λ (11). The repetitive sequence within the *stf* protein gene encoding the λ tail fiber was subcloned into plasmid expression vectors by the polymerase chain reaction (PCR). The expression of the target DNAs was carried out in *E. Coli* BL21(DE3) induced by Isopropyl-β-thiogalactoside (IPTG) (1mM). The harvested proteins were purified with an immobilized metal ion affinity chromatography loaded with Ni^{2+}. The desired proteins—periodic polypeptides containing up to $x = 32$ repeats of the septapeptide sequence $[SAxxAxx]_x$ (**1**)—were then characterized by gel electrophoresis, mass spectroscopy, and circular dichroism (CD). Our ultimate goal is to self assemble monodisperse helices in an unidirectionally oriented array as one step in the fabrication of polymer thin films wherein molecular-engineered functionalities (chemical, electrical, or nonlinear-optically-active) may be introduced into strata coplanar with a substrate.

EXPERIMENTAL PROCEDURES

Construction of Expression Plasmids

The sequences corresponding to aminoacids 123-262, 123-290, 123-318, and 123-346, which are designated as 420, 504, 588, and 672 respectively according to their gene fragment size, of the mature λ *stf* protein were amplified by the polymerase chain reaction (PCR) using gene-specific primers and cloned into the NdeI/XhoI sites of the pET-16b vector (Novagen), upstream of the 10 His•Tag sequence. Primers for PCR were designed to contain between 18 and 20 nucleotides of gene-specific sequence as well as recognition sequences for restriction enzymes to facilitate cloning. 3' Primers contained translation termination codons to prevent read-through translation into plasmid sequences. PCR reactions contained 50 ng linearized template plasmid, 200 μM each dNTP, 50 pmoles of each 5' and 3' primer, in 20 mM Tris-HCl, pH 8.75, 10 mM KCl, 10 mM $(NH_4)_2SO_4$, 2 mM $MgCl_2$, 0.1% Triton X-100, 0.1 mg/ml BSA, and 2.5 U *Pfu*

(Stratagene) or 1.0 U*AmpliTaq* (Perkin-Elmer) DNA polymerase. Reactions were cycled 30 times; 1 min. at 94°C, 1 min. at 55°C, and 1 min. at 72°C. After a final extension at 72°C for 10 min. reaction products were collected, digested with NdeI and XhoI, purified by agarose gel electrophoresis and cloned into the NdeI/XhoI sites of plasmid pET-16b. Plasmids were maintained and characterized in *E. coli* strain DH5α.

Protein Expression and Purification

Recombinant plasmids were transferred to host strain BL21(DE3) which contained a chromosomal copy of the gene for T7 RNA polymerase and grown at 37 °C. Expression was induced by adding IPTG to a concentration of 1 mM. Induction was continued for 2 h. Cells were collected by centrifugation at 5,000 x g for 5 min. and lysed by sonication. Each expressed peptide was purified by immobilized metal ion affinity chromatography on chelating sepharose loaded with Ni^{2+} (Novagen) according to the protocols provided by the manufacturer. Recombinant *stf*-His•Tag peptides were recovered by elution with 1 M imidazole. Recombinant peptide was soluble within the bacterial cell and therefore could be purified under mild, non-denaturing conditions. Imidazole was removed by dialysis and the final preparations were concentrated by ultrafiltration using Amicon Centriprep-3 concentrators.

Mass Spectrometry

Molecular masses of the expressed peptide chains were determined with matrix-assisted laser desorption/ionization (MALDI) mass spectrometry (MS) by Dr. Paul Danis (Rohm & Has). The MALDI MS was performed on a Bruker REFLEX time-of-flight mass spectrometer using sample preparation and analysis procedures similar to those in reference (12).

Circular Dichroic Spectrophotometry

Circular dichroic spectra were recorded with a ultraviolet CD spectrophotometer (AVIV Model 60DS). Part of the peptide stock solution in a quartz cuvette (Helma, 1-mm pathlength) was diluted with 25 mM phosphate buffer (pH 7.5) until the UV absorbance was between 0.4 and 1.0, near the value of 0.869 for the best ratio of signal to noise (13). Peptide concentration was determined using bovine serum albumin (BSA) standard solution as a reference. The ellipticity (milidegrees) was scanned 4 to 5 times from 190 to 260 nm at 0.5-nm intervals. Data points at each wavelength were subtracted from the spectrum of the phosphate buffer, averaged, and converted from ellipticity to mean molar ellipticity per residue ($[\theta]$, deg.cm²/dmol) by multiplying by Cb/n, where C was the concentration of the peptide, b was the pathlength of the cell (1 mm), and n was the number of residues in the molecules. The resulting data points were subjected to a five point smoothing algorithm and graphed. The α-helicity of a peptide in solution was estimated from the ratio of molar ellipticities $[\theta]_{222}/[\theta]_{max}$, where $[\theta]_{222}$ was the observed mean ellipticity per residue at 222 nm and $[\theta]_{max}$ was the maximum theoretical ellipticity at 222 nm; the latter was calculated as -39,500x(1-2.57/n) deg cm^2/dmol where n is the number of residues per polypeptide(14).

RESULTS AND DISCUSSION

Gene Construction

The strategy used to prepare naturally occurring proteins of repetitive septad sequences of structure **1** is outlined in Figure 1. Regions of the λ *stf* gene were obtained by the polymerase chain reaction employing gene-specific primers flanking the regions of interest. A single 5' PCR primer was used in conjunction with one of four different 3' PCR primers to amplify DNA fragments of different sizes encoding varying lengths of the repeat sequences. PCR products were cloned into the vector, pET-16b, by virtue of restriction endonuclease sites included in the design of the PCR primers. The resulting recombinant plasmids were transferred into the *E. coli* strain BL21(DE3) and peptide production was induced by the addition of IPTG.

Four different chain length gene fragments 420, 504, 588, and 672 of the mature λ *stf* protein were amplified by the polymerase chain reaction using gene-specific primers. Primers for PCR were designed to contain between 18 and 20 nucleotides of gene-specific sequence as well as recognition sequences for restriction enzymes to facilitate cloning. 3' Primers contained translation termination codons to prevent read-through translation into plasmid sequences. The sequences of 5' and 3' DNA primers are shown in Table 1.

The products from PCR were characterized by agarose gel electrophoresis (Figure 2). Eighteen μl of product from each polymerase chain reaction was resolved on a 2% agarose gel and stained with ethidium bromide. Lane M, contains DNA size markers. Lanes 1 through 4 contain the products from PCR with the single 5' PCR primer and 3' PCR primers 1, 2, 3, and 4. The size of the products are 672 bp, 588 bp, 504 bp and 420 bp respectively.

Protein Expression.

The host used for protein expression was *E. Coli* strain BL21-(DE3) which contains a chromosomal copy of the gene for T7 RNA polymerase. The host is a lysogene of bacteriophage DE3, a lambda

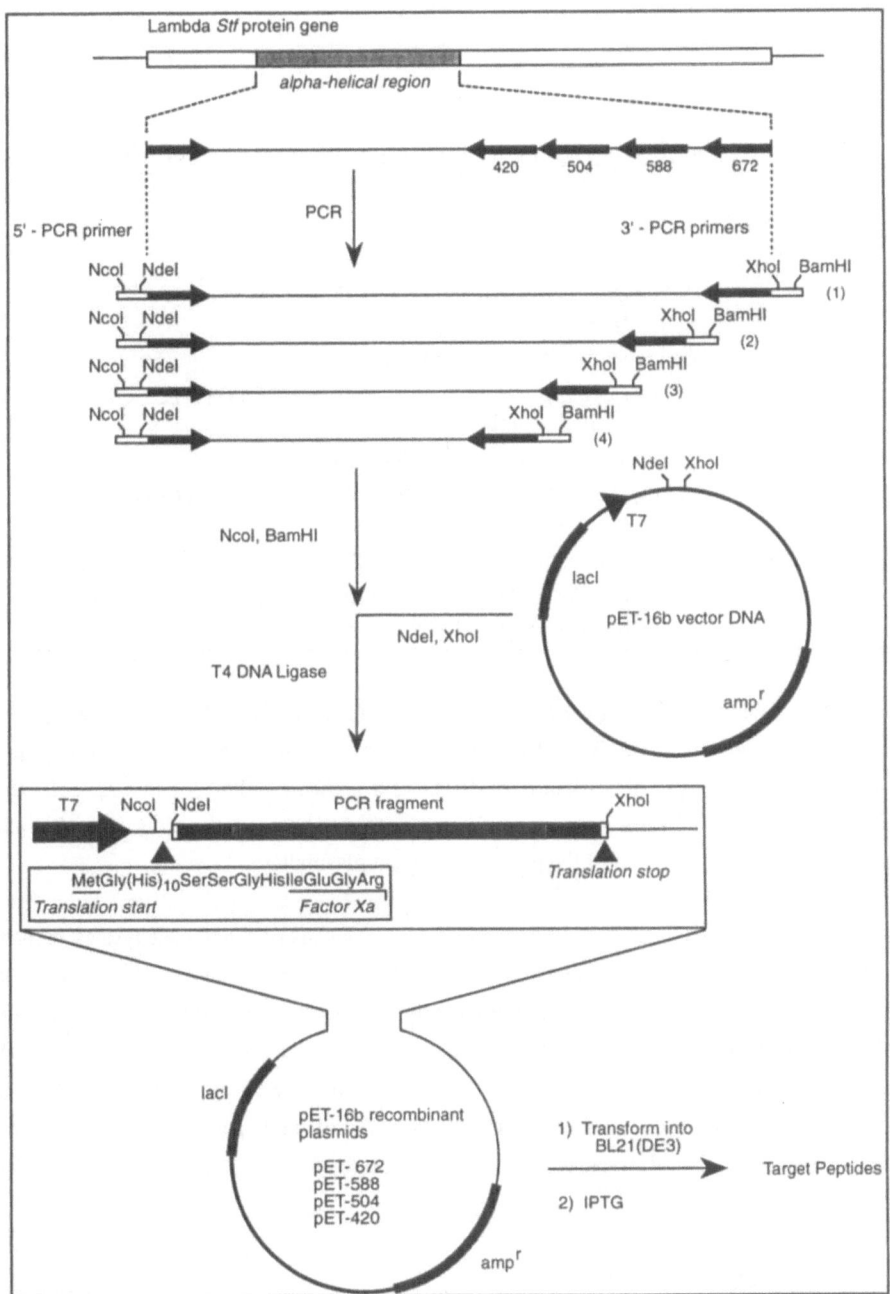

Figure 1. Strategy for the cloning and expression of different length λ *stf* gene fragments encoding repetitive amino acid sequence.

Table 1. The DNA primers for polymerase chain reaction of the *stf* protein gene encoding the λ tail fiber.

5' Primer:

    (corresponding to aminoacids 123-128)

    5'-A<u>CCATGG</u>CCC<u>ATATG</u>AGTACGGCAGACGCGAAG-3'
        NcoI   NdeI

3' Primers:

  (1).   (corresponding to aminoacids 340-346)

        5'-A<u>GGATCC</u><u>TCGAG</u>**TCA**ATCTTCTGCACGTTTTGCCG-3'
           BamHI XhoI TERM

  (2).   (corresponding to aminoacids 313-318)

        5'-A<u>GGATCC</u><u>TCGAG</u>**TCA**TCCCGCAGCCTCTGTCGCC-3'
           BamHI XhoI TERM

  (3).   (corresponding to aminoacids 284-290)

        5'-A<u>GGATCC</u><u>TCGAG</u>**TCA**CCGTTCCGCTGCTGTTTC-3'
           BamHI XhoI TERM

  (4).   (corresponding to aminoacids 257-262)

        5'-A<u>GGATCC</u><u>TCGAG</u>**TCA**GGAAGCTGCACGACCGGCAC-3'
           BamHI XhoI TERM

derivative that contains the gene for T7 RNA polymerase (15). The T7 RNA polymerase gene is under control of the *lacUV5* promotor, which is inducible by isopropyl-β-D-thiogalactopyranoside (IPTG). Addition of IPTG to a growing culture of this *E. coli* strain induces expression of T7 RNA polymerase, which in turn transcribes the cloned λ *stf* peptide sequences from the T7 promoter of the pET-recombinant plasmids.

Protein expression was characterized by electrophoretic analysis of whole-cell lysates prepared immediately before or 2 h after induction with IPTG. Prominent bands for all four target proteins were apparent after 2 h, whereas these bands were barely visible prior to induction (data not shown). In each case, the new product migrated as a single band and no evidence was found for the accumulation of chain length variants that could have arisen from polymorphism in the insert DNA or from premature translational termination. All of the target proteins were found in the soluble fraction of the cell lysate (data not shown); none appeared to accumulate in inclusion bodies.

Protein Purification and Analysis

For purification of expressed peptides, the vector supplied a His•Tag sequence which contained a stretch of 10 histidine residues that were expressed at the N-terminal end of each of the target proteins. These His•Tagged proteins produced in bacterial cells could therefore be purified by chromatography on a Ni^{2+}-NTA column (Novagen). SDS-PAGE analysis demonstrated that all four peptides were retained on the Ni^{2+}-NTA column at pH 7.9 after extensive washing with a Tris-HCl buffer containing 0.5M NaCl, and 60 mM imidazole. Non-specific bacterial proteins were eluted from the column in the flow-through material. The target peptides with leading His•Tag sequence were eluted from the column with 1 M imidazole. The purity of the final polypeptides was judged to be greater than 95% by SDS-PAGE (Figure 3). Typical yields after purification were ca. 10 mg/L of culture.

The expressed polypeptides are also designated as 420, 504, 588, and 672 respectively and the amino acid sequence of these proteins is given in sequence 2:

<u>[M]G(HHHHHHHHHH)SSGHIEGRHM</u>STADAKKSAGDASASAAQVAALVTD-
　　　　His•Tag sequence

ATDSARAASTSAGQAASSAQEASSGAEAASAKATEAEKSAAAAESSKNAAAT-

SAGAAKTSETNAAASQQSAATSASTAATKASEAATSARDAVASKEAAKSSET-

NASSSAGRAAS$_\Delta$SATAAENSARAAKTSETNARSSETAAER$_\Delta$SASAAADAKTA-
end of peptide **420** (15,327 Da)　　　　　　　　end of peptide **504** (18,146 Da)

AAGSASTASTKATEAAG$_\Delta$SAVSASQSKSAAEAAAIRAKNSAKRAED$_\Delta$
　　end of peptide **588** (20,522 Da)　　　　　　end of peptide **672** (23,280 Da)

2

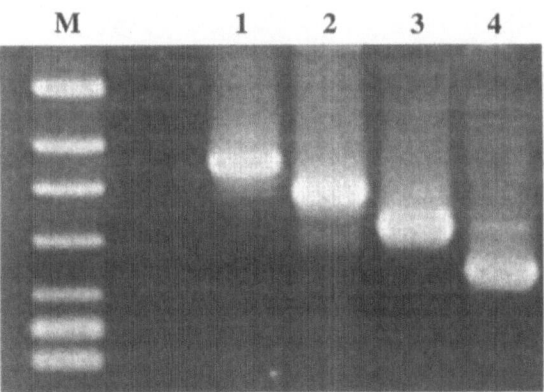

Figure 2. Gel electrophoresis of PCR products. Eighteen µl of product from each polymerase chain reaction was resolved on a 2% agarose gel and stained with ethidium bromide. Lane M, contains DNA size markers. Lanes 1 through 4 contain the products from PCR with the single 5' PCR primer and 3' PCR primers 1, 2, 3, and 4.

We observe with matrix-assisted laser desorption/ionization (MALDI) mass spectrometry (courtesy of Dr. Danis of Rohm and Hass company), molecular ions of the four target proteins at m/z values in good agreement with those expected; the differences between measured and expected values are less than 50 Da. The mass spectra of the polypeptides show only one peak for each protein in the mass range of 12,000 and 29,000 and indicate that the expressed polypeptides are pure and monodisperse.

Helicity of the Polypeptides.

The CD analysis efforts were focused on the two longer polypeptides 672 and 588, containing x=32 and x=28 repeats of the septapeptide sequence **1** of the λ tail fiber respectively. The CD spectra

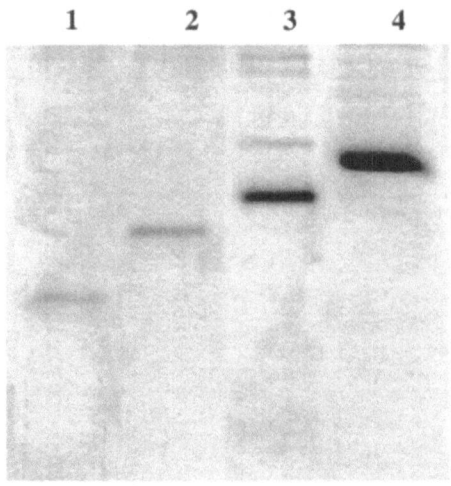

Figure 3. Gel electrophoresis of four purified polypeptides. Polypeptides were purified by Ni^{2+}-chelating chromatography and analyzed by SDS-PAGE on 15% gels. The gel was stained with coomassie brillient blue and photographed. Lanes 1 through 4 represent peptide products from DNA fragments 420, 504, 588 and 672 respectively.

of polypeptides 672 and 588 with His•Tag were similar at pH 7.5 and 25 °C (Figure 5). Both spectra were bimodal with large minima at 206-208 nm and 222 nm, indicating a significant proportion of the residues in the α-helical conformation (16). $[\theta]_{222}/[\theta]_{208}$ is about 0.8 for single-stranded α-helix and about 1.0 for a double-stranded α-helical coiled coil (17). By this criterion, both polypeptides 672 and 588, with $[\theta]_{222}/[\theta]_{208} \sim 0.8$ and 0.75 respectively, each contain a single-stranded α-helix at pH 7.5 and 25 °C. The conformation was very stable for temperatures between 0 °C and 65 °C. The histidine tag was removed from the polypeptides by resuspending the proteins

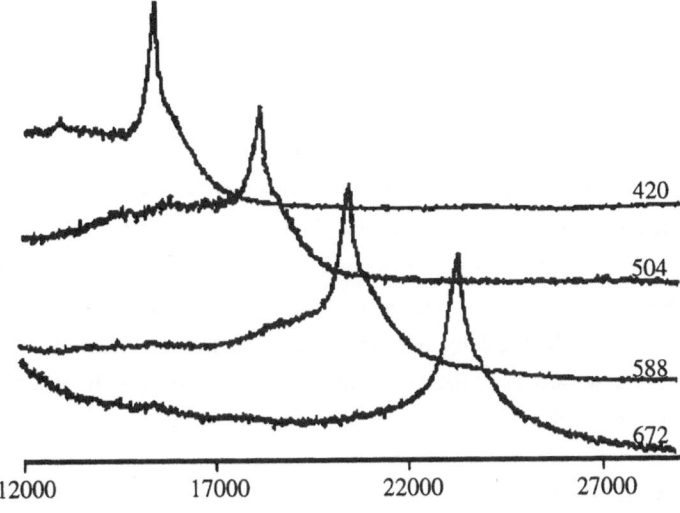

Figure 4. Matrix-assisted laser desorption/ionization time-of-flight mass spectra for the expressed proteins.

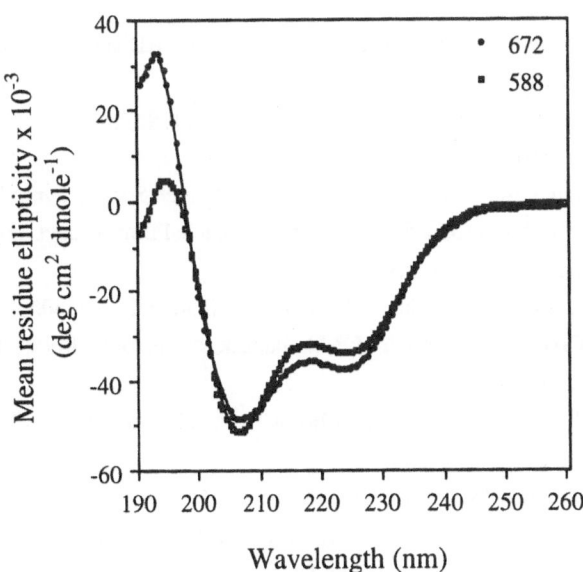

Figure 5. Circular dichroic spectra of polypeptides in 25 mM phosphate buffer (pH 7.5) at 25 °C: (•) polypeptide 672; (□) polypeptide 588.

in an aqueous solution containing Factor Xa protease. Detailed analysis of the solution and solid-state structure of these tag-free polypeptides are in progress.

CONCLUSIONS

Four rodlike naturally occurring polypeptide sequences were identified and expressed in *E. coli*. All of the four polypeptides were genetically stable in the expression system used. MALDI mass spectrometry confirms the sizes of the proteins which were obtained in a yield of approximately 10 mg from 1-L culture. The preliminary CD studies reveal that the two longer polypeptides with their histidine tags are each comprised of a single-stranded α-helix at pH 7.5 and 25 °C. This is a relevant finding as we anticipate that the septapeptide [SAxxAxx]$_x$ amino acid sequence in the (tag-free) proteins should show a tendency to adopt coiled coil quaternary structures in the tail fibers (18). Generally speaking our results demonstrate the feasibility of manipulating biological processes to produce macromolecular products that may be used in thin film applications in materials science.

ACKNOWLEDGMENTS

This work was funded in part by a grants from the NC Biotech Center, Grant # 9313-ARG-0904 and NIH, Grant # DK42701. We thank Cecelia H. Branan for help with the protein harvesting experiments and R.W. Hendrix for making the λ *stf* gene available to us.

REFERENCES

1 Enriquez E, Gray KH, Guarisco VF, Linton RW, Mar KD and Samulski ET (1992) J Vac Sci Technol A 10:2775
2 Enriquez E, Jin MY, Jarnagin RC and Samulski ET (1993) Materials Research Society Symp Proc 292:163
3 Enriquez E and Samulski ET (1992) Materials Research Society Symp Proc 225:423
4 Ulman A (1991) An Introduction to Ultrathin Organic Films: From Langmuir-Blodgett to Self-Assembly, Academic Press: Boston
5 Worley CG, Linton RE and Samulski ET (1995) Langmuir, 11:3805
6 Cappello J, Crissman J, Dorman M, Mikolajczak M, Textor G, Marquet M, and Ferrari F (1990) Biotechnol Prog 6:198
7 McGrath KP, Tirrell DA, Kawai M, Mason TL, and Fournier MJ (1990) Biotechnol Prog 6:198
8 Parkhe AD, Fournier MJ, Mason TL and Tirrell DA (1993) Macromolecules 26:6691
9 Mahajan SK (1988) In: Kucherlapati R, Smith GR (eds) Genetic Recombination. American Society for Microbiology, Washington DC
10 Zhou NE, Zhu BY, Kay CM, and Hodges RS (1992) Biopolymers 32:419
11 Hendrix RW, and Duda RL (1992) Science 258:1145
12 Danis PO, Karr DE, Westmoreland DG, Piton MC, Christie DI, Clay PA, Kable SH, and Gilbert RG (1993) Macromolecules 26:6684

13 Johnson WC (1988) Annu Rev Biophys Chem 17:145

14 Chen YH, Yang JT and Merrifield RB (1981) Biochemistry 13:3350

15 Studier FW, and Moffatt BA (1986) J Mol Biol 189:113

16 Holzwarth G, and Doty P (1965) J Am Chem Soc 87:218

17 Lau SYM, Taneja AK, and Hodges RS (1984) J Biol Chem 259:13253

18 Cohen CC and Parry DAD (1994) Science 263:488.

Enzymatic Synthesis of Polysaccharides: A New Concept in Polymerization Chemistry

S. Kobayashi, S. Shoda

Department of Materials Chemistry, Graduate School of Engineering
Tohoku University, Sendai 980-77, Japan

Abstract: Various poly- and oligosaccharides such as cellulose, maltooligosaccharides, alternating 6-O-methylcellulose, and xylan have been prepared by the enzymatic polymerization of glycosyl fluoride monomers catalyzed by glycosidases like cellulase and amylase with perfect regioselectivity and stereoselectivity.   By using the new architectural methodology of enzymatic polymerization technique, the metastable cellulose I with anti-parallel glucan chain structure was constructed *in vitro* for the first time.   Based on these findings, a new concept of "choroselectivity" expressing the spacial direction control in ordering of macromolecular chains during polymerization has been proposed.

INTRODUCTION

Polysaccharides are among the most important biopolymers as are proteins and nucleic acids in nature.   Polysaccharides are high molecular weight carbohydrates formed as a result of condensation reaction with elimination of water between the hydroxy group at the C-1 carbon atom of a monosaccharide unit and a hydroxy group of another monosaccharide unit. Recent progress for developments of selective synthesis and polymer synthesis has made it possible to produce various useful polysaccharide derivatives in material science, pharmaceutical science, the food industry etc.   Most of the synthetic approaches so far reported have, however, been based on principle of chemical modification of a naturally occurring polysaccharide.   Few synthesis of polysaccharides starting from a mono or disaccharide building block have been reported, probably due to the following reasons: 1) In spite of progress on the polycondensation reactions of partially protected sugar derivatives [1] and ring-opening polymerization of anhydrosugars [2], complete stereocontrol of the glycoside bond-forming reactions has not been achieved.   2) Chemical synthesis of polysaccharides requires complicated procedures including a regioselective blocking and deblocking of a hydroxy group.   Here, we wish to propose a new principle for the architecture of a polysaccharide skeleton starting from a monosaccharide or disaccharide monomer as building block by using an enzyme as catalyst.   The use of the enzyme for the

M. Kamachi · A. Nakamura (Eds)
New Macromolecular Architecture and Functions
Proceedings of the OUMS '95 Toyonaka, Osaka, Japan, 2-5 June, 1995
© Springer-Verlag Berlin Heidelberg 1996

glycosylation process is considered to be the most promising way for construction of a glycosidic linkage under perfect control of regio- and stereochemistry.

SYNTHETIC CELLULOSE

Enzymatic Polymerization of β-Cellobiosyl Fluoride

Cellulose, a representative of polysaccharides, is the most abundant organic polymer occurring on earth.   The utilization of cellulose or derivatized cellulose as functional material has so far been based on natural cellulose or modification of natural cellulose via a polymer reaction such as alkylation, nitration and esterification.   Despite enormous effort devoted to cellulose synthesis *in vitro* [3,4,5],   there have been no previous reports of a chemical process producing a cellulose backbone in an artificial system.

Recently, however, we have achieved the first chemical synthesis of cellulose by a polycondensation of β-cellobiosyl fluoride substrate monomer catalyzed with cellulase, an extracellular hydrolysis enzyme of cellulose, in an organic solvent/buffer mixed medium [6].

synthetic cellulose

The effect of the monomer structure on the polymerization was investigated using various cellobiose derivatives.   α-Cellobiosyl fluoride whose fluoride atom at the anomeric position has the opposite configuration did not polymerize at all.   This result shows that the anomeric stereochemistry plays an important role in this enzymatic polymerization.   Other cellobiose derivatives, alkyl cellobioside and 1-*O*-acylcellobiose, alkyl thiocellobioside, and cellobiosyl sulfoxide were found to be recognized by the cellulase. These cellobiosyl monomers, however, showed poor polymerizability in comparison with that of β-cellobiosyl fluoride monomer [7].

β-Cellotriosyl fluoride and β-cellotetraosyl fluoride were prepared as novel substrate monomers, and their behaviors of hydrolysis and polymerization by cellulase were investigated. These substrate monomers were found to be rapidly hydrolyzed in aqueous solution, indicating that they can be recognized as a substrate by the cellulase. A polymerization reaction was performed in an aqueous organic solvent, giving rise to cellooligomers and cellulose [8].

The formation of the stereoregular polysaccharides may be explained by assuming the following two processes. The first step involves the formation of a glycosyl-enzyme intermediate or a glycosyl oxocarbenium ion at the active site of cellulase with the elimination of fluorine atom. This reactive intermediate is then attacked by the 4'-hydroxy group of another monomer which locates in a subsite of the enzyme, leading to the stereoselective formation of the β(1→4) linkage. Consequently, the stereochemistry of the product is retention of configuration via double inversion concerning the anomeric carbon atom of the β-cellobiosyl fluoride.

The present reaction mechanism is to be interestingly compared with a biosynthetic pathway of cellulose that involves the inversion of configuration concerning the C1 carbon atom of the substrate of uridine diphosphate-glucose (UDP-glucose) [9].

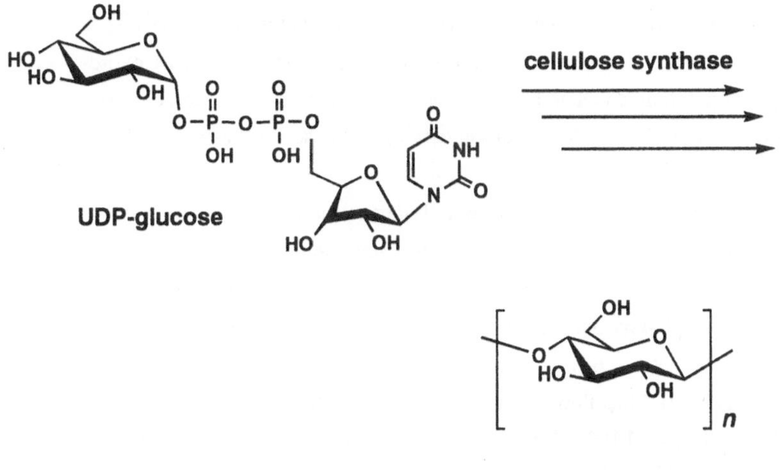

cellulose synthase

UDP-glucose

natural cellulose

Visualization of Cellulose Forming Process and Morphology of Synthetic Cellulose

According to the enzymatic polymerization, it is possible to form an elongated polymer chain of cellulose in solution, enabling visualization of cellulose-forming process and of the subsequent crystallization process. These processes were visualized by transmission electron microscopy. The allomorph of the synthetic cellulose was the crystalline cellulose II, a thermodynamically more stable form with an anti-parallel glucan orientation, which was verified by means of electron diffraction of the synthetic cellulose [10].

Further studies on the polymerization conditions, eg, purification of the enzyme, the solvent composition etc, led to the production of native type cellulose I, a metastable allomorph [11]. The production of microfibrillar cellulose I results from micellar aggregations of the cellulase providing an appropriate environment for the production of unidirectionally ordered glucan chains with the same polarity and an extended-chain conformation. Cellulose I is one of nature's most perfect biopolymers with a parallel glucan chain orientation and has long been believed to be impossible to be prepared *in vitro*. The present experiment is the first *in vitro* synthesis of cellulose I via a non-biosynthetic pathway using enzymatic polymerization and may provide a new insight for the elucidation of cellulose synthesis in nature.

CHOROSELECTIVITY

In the preparation of synthetic cellulose, three categories of selectivity were achieved by using an enzyme as catalyst. They are chemoselectivity (exclusive formation of the glycosidic linkage), regioselectivity (selective intermolecular ether bond formation between 1 and 4 position), and stereoselectivity (selective formation of glycosidic bond with β orientation at the C1 atom). All of these selectivities are defined as expressions of the selectivities concerning covalent bond formations between two reaction centers, the anomeric carbon atom and the 4-hydroxy group, in case of cellulose synthesis.

The cellulose I formation means that the polymerization is controlled not only by the above three selectivities but also by a selectivity concerning the relative direction of the glucan chain ordering. This is, to our best knowledge, the first successful example of controlling the relative direction of polymer chains at the polymerization stage. For understanding this new findings, we wish to propose a new concept of "choroselectivity" concerning the spacial (three dimensional) direction control in ordering of macromolecular chains during polymerization [12,13,14]. The term "choros" has its origin in a Greek word which means "space".

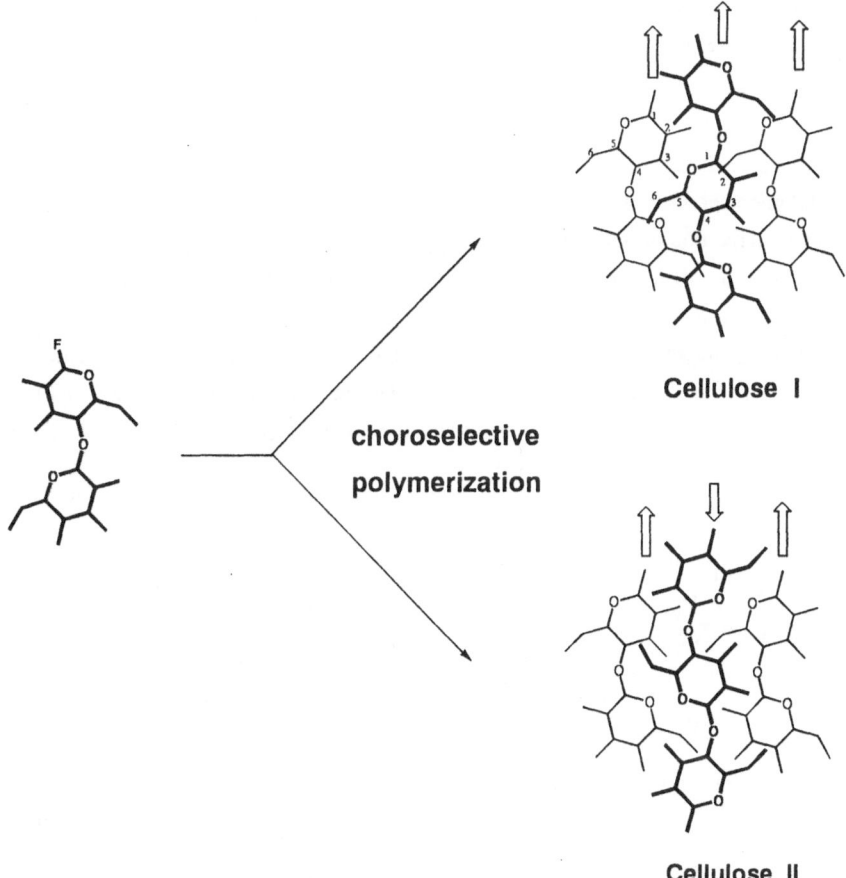

Cellulose I

choroselective
polymerization

Cellulose II

This new concept is widely applicable to all polymerization reactions which produce polymer chains with spacial direction. Such three dimensional control, that is choroselective polymerization, is a crucial consideration problem when synthetic chemists design a polymerization reaction considering not only the polymer's unit structure, sequence, stereochemistry, molecular weight and its distribution, end-group structure etc, but also its physical properties as a functional material.

ENZYMATIC SYNTHESIS OF MALTOOLIGOSACCHARIDES AND RELATED
OLIGOSACCHARIDE

The enzymatic polycondensation of a glycosyl fluoride monomer in aqueous organic solvent
media has successfully been applied to the amylase-catalyzed preparation of
maltooligosaccharides [15]. Generally, maltooligosaccharides are produced by a
degradation reaction of polymers such as amylose, amylopectin, and glycogen, whereas only
few studies on their production from monomers, e.g. the condensation of a glucose
derivative or a maltose derivative, have been reported so far. The polymerization was
carried out by stirring a mixture of α-maltosyl fluoride and α-amylase in an organic solvent/
phosphate buffer at room temperature. The polycondensation reaction took place
effectively when the reaction was carried out in a methanol/buffer (2:1) mixed solvent
system.

maltooligosaccharides

An unnatural oligosaccharide having an α(1→4) and α(1→6) linkage alternatingly has also
been prepared using α-maltosyl fluoride as substrate for pullulanase, a hydrolysis enzyme
which cleaves the α(1→6) glycosidic bond in pullulan [16].

MACROMOLECULAR ARCHITECTURE OF MODIFIED CELLULOSE

New cellobiosyl fluoride derivatives, 6-*O*-methyl and 6'-*O*-methyl-β-cellobiosyl fluorides were prepared as monomers for enzymatic polymerization [17]. In the hydrolytic behavior of these monomers, both were hydrolyzed by the action of cellulase to give the corresponding 1-hydroxy sugar derivatives. These results show that the monomers can be recognized as a substrate by the cellulase and are capable to form an enzyme-substrate complex. Enzymatic polymerization of 6-*O*-methylated monomer took place smoothly in an aqueous organic solvent (CH$_3$CN/acetate buffer) giving rise to a novel cellulose derivative having a methyl group alternatingly at the 6 position. The ^{1}H and ^{13}C NMR spectra of the resulting products indicated that the enzymatic glycosylation process is under perfect control of regio- and stereochemistry, leading to the β(1→4)glycosidic linkage. In contrast, 6'-*O*-methylated monomer showed a lower polymerizability. It is suggested that the difference of polymerizability is explained by the steric repulsion between the methyl group of the monomers and the amino acid residues in the catalytic site of the cellulase [18].

alternating 6-O-methyl cellulose

alternating 6-O-methyl cellooligomer

SYNTHETIC XYLAN

Synthetic xylan, a xylose polymer composed exclusively of β(1→4) glycosidic bonds, has been synthesized by the enzymatic polymerization of β-xylobiosyl fluoride monomer catalyzed by cellulase (*Trichoderma viride*) [19]. Xylan is one of the most important components of hemicellulose in plant cell wall. Naturally occurring xylan normally contains L-arabinose and 4-*O*-methylglucuronic acid as minor unit in the side chain. The synthesis of polysaccharides consisting exclusively of a xylopyranose moiety connected through a β(1→4) glycosidic bond has, therefore, been a challenging problem in the synthetic field of polysaccharide chemistry.

Structure of Natural Xylan

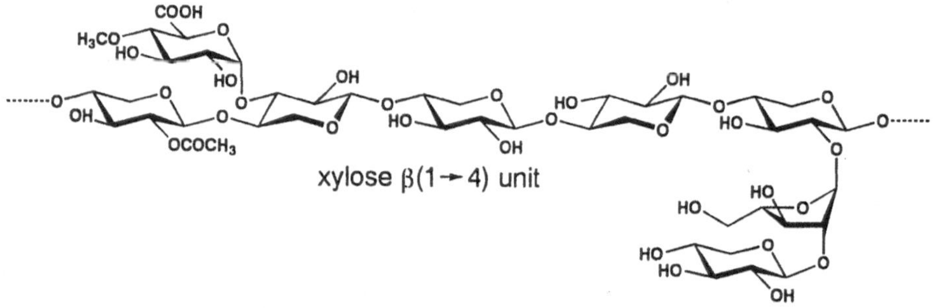

xylose β(1→4) unit

When the β-xylobiosyl fluoride monomer was treated with a catalytic amount of cellulase in acetonitrile-buffer system, the substrate monomer polymerized smoothly to afford the corresponding polycondensation products.

The cross-polarization/magic angle spinning (CP/MAS) solid ^{13}C NMR spectrum of the water-insoluble product shows signals which are assignable to all carbon atoms in xylopyranose moiety with β(1→4) linkage. No signals derived from a β(1→3) glycosidic bond were observed in the products, indicating that only products with β(1→4) linkages were formed. In order to determine the molecular weight of the synthetic xylan, it was converted to the corresponding carboxymethyl derivative using chloroacetic acid/NaOH. The gel permeation chromatographic (GPC) analysis of the carboxymethylated product revealed that the number average molecular weight is at least 6. 7 x 10^3 which corresponds to the degree of polymerization, DP = 23.

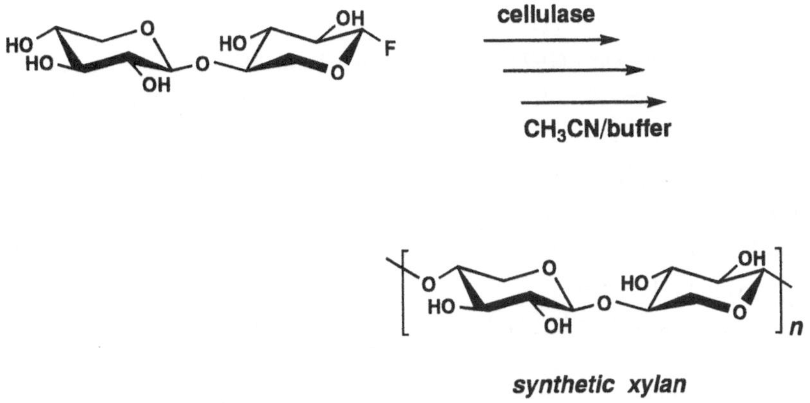

synthetic xylan

It is to be noted that unlike natural xylan the synthetic xylan prepared by the enzymatic polymerization of β-xylobiosyl fluoride consists exclusively of xylopyranose residues with neither L-arabinose nor glucuronic acid as side chain.

CONCLUSION

Various important polysaccharides have successfully been synthesized *in vitro* by the enzymatic polymerization of glycosyl fluoride monomers. The formation of native cellulose I *in vitro* has been realized for the first time by using this novel methodology. Based on these findings, a new concept "choroselectivity" in polymerization chemistry was proposed. The present mew methodology for polysaccharide synthesis is a promising tool for preparation of highly designed polysaccharide derivatives, and continues to broaden the scope of macromolecular architecture by a biocatalyst in future.

References:

1 Kochetkov NK   (1987)   Tetrahedron 43: 2389

2 Schuerch C   (1972)   Adv. Polym. Sci   10: 173

3 Husemann E,   Müller GJ   (1985)   Makromol. Chem. 91: 212

4 Micheel F,   Brodde OE   (1974)   Liebigs Ann. Chem. 702

5 Uryu T, Yamaguchi C, Morikawa K, Terui K, Kanai T, Matsuzaki K   (1985)   Macromolecules   18: 599

6 Kobayashi S,   Kashiwa K,   Kawasaki T,   Shoda S   (1991)   J. Am. Chem. Soc. 113: 3079

7 Shoda S,   Shimada J,   Wen X,   Obata K,   Okamoto E,   Kobayashi S   (1992)   Polym. Prepr. Jpn   41: 2418

8 Osada S,   Okamoto E,   Shoda S,   Kobayashi S   (1995)   Polym. Prepr. Jpn   44: 2468

9 Delmer DY   (1983)   Adv. Carbohydr. Chem. Biochem.   41: 105

10 Kobayashi S, Shoda S,   Lee JH,   Okuda K,   Brown RM Jr,   Kuga S   (1994)   Mcromol. Chem. Phys.   19:1319

11 Lee JH,   Brown RM Jr,   Kuga S,   Shoda S, Kobayashi S   (1994)   Proc. Natl. Acad. Sci. USA   91: 7425

12 Kobayashi S,   Shoda S   (1995)   Int. J. Biol. Macromol.   17:373

13 Shoda S,   Kobayashi S   (1995)   Macromol. Symp. 99:179

14 Kobayashi S,   Okamoto E,   Wen X,   Shoda S   J. Macromol. Sci.-Pure Appl. Chem. in press

15 Kobayashi S,   Shimada J,   Kashiwa K, Shoda   S   (1992)   Macromolecules 25:3237

16 Kobayashi S,   Shimada J, Wen X,   Shoda S   (1991)   Polym. Prepr. Jpn 40:3035

17 Shoda S, Okamoto E, Kiyosada T,   Kobayashi S   (1994)   Macromol. Rapid Commun. 15: 751

18 Okamoto E,   Shoda S,   Kobayashi S   (1995)   Polym. Prepr. Jpn.   44:2305

19 Kobayashi S,   Wen X,   Shoda S   Macromolecules   in press

Self Assembling Organic Nanotubes

Jeffrey D. Hartgerink, M. Reza Ghadiri

The Scripps Research Institute
Departments of Chemistry and Molecular Biology
10666 N. Torrey Pines Rd.
La Jolla, CA 92122 USA

Abstract: Traditionally microscopic devices have been made by being cut or formed from larger objects, but as the dimensions of these products shrink below the micron level this process becomes increasingly difficult. Recently chemists have begun to try the opposite approach, that is building these nanoscale objects from molecular building blocks. Although these devices are too small to be manufactured by traditional materials science approaches, they are also far too large to be synthesized by classical chemical synthesis. In order to reach these nanoscale devices from a molecular level up, a massively convergent synthesis is required. Production of these nanoscale objects however, is not unknown and has been occurring for over three billion years in living biological systems. From microtubules to viruses, nature uses a broad variety of self-assembly techniques to build its sub-cellular machines that ultimately lead to life. Taking lessons from nature, we have designed and synthesized a number of self-assembling peptide based nanotubes. Nanotubes in general have recently attracted much attention, especially from carbon based nanotubes. Speculation on their utility range from molecular wires to catalysts to novel drug delivery systems. Peptide based nanotubes have many attractive features that make them a particularly good system to address these possibilities. For example, nanotubes based on cyclic peptides have easily alterable surface chemistries, adjustable internal diameters, open ends for including compounds such as metal or passing through ions and small molecules, and are also easily made by combining self-assembly with standard solid phase peptide synthesis. Furthermore, cyclic peptide based nanotubes have already been found to function as ion and small molecule membrane channels.

INTRODUCTION:

Nanotubes have recently received a great deal of attention and have been prepared from a variety of materials and taken a variety of forms including polymerized cyclodextrins[1], silica cast cylindrical micelles[2], all inorganic zeolites[3] as well as all carbon multilaminar cylinders[4]. These nanotubular objects are of interest both because of their novel architecture and their potential utility.

Our approach to the construction of nanotubular objects is based on cyclic peptides. In 1974 De Santis et al., in a theoretical conformational analysis of linear poly D-,L-peptides had suggested the possibility of forming cylindrical structures from cyclic peptides[5]. However,

M. Kamachi · A. Nakamura (Eds)
New Macromolecular Architecture and Functions
Proceedings of the OUMS '95 Toyonaka, Osaka, Japan, 2-5 June, 1995
© Springer-Verlag Berlin Heidelberg 1996

early attempts at the design of peptide tubular ensembles based on those principles were unsuccessful as evidenced by X-ray crystal structure analysis[6]. According to our design principles[7], cyclic peptide structures made up of an even number of alternating D- and L-amino acid residues, tend to minimize nonbonded intramolecular transannular side chain-side chain and side chain-backbone interactions by adopting (or sampling) a flat ring-shaped conformation in which all backbone amide functionalities lie approximately perpendicular to the plane of the ring structure. Furthermore, because of the local conformational and steric constraints imposed by the alternating amino acid backbone configuration, all amino acid side chains point outward away from the center of the peptide ring structure thus leaving it free to form a tubular core structure. Once in this flat conformation, the amide backbone of the cyclic peptide may form an extensive hydrogen bonding network perpendicular to the plane of the peptide. This β sheet hydrogen bonding drives the self assembly process along the z axis while hydrogen bonding and hydrophobic packing drive the assembly of bundles of nanotubes.

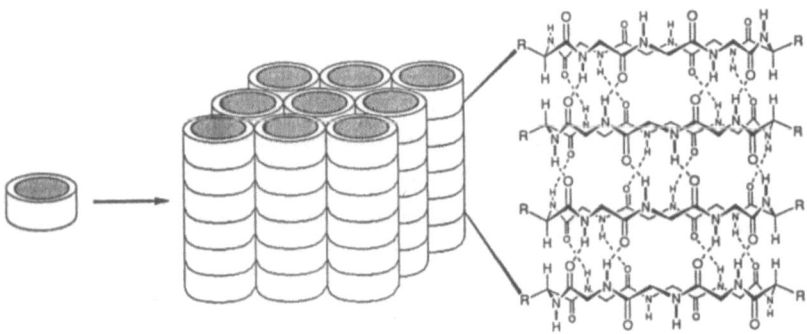

Figure 1) Illustration of the hydrogen bonding network formed upon self-assembly in addition to the parallel association of nine individual nanotubes.

SOLID STATE NANOTUBES

One of the advantages of using cyclic peptides as the building blocks for nanotubes is their ease of synthesis. Cyclic peptides can be prepared by a procedure based on the optimized BOC chemistry of Kent and coworkers[8] and the cyclization technique of Rovero et al.[9]. These methods allows rapid, pure, high yield synthesis of cyclic peptides. Because the building blocks are peptide based, the chemistry and diameter of the building block - and thus the resulting nanotube - may be easily altered.

The conditions under which the cyclic peptides self assemble to form nanotubes is dependent on the type of amino acids which make up the cyclic peptide. The first nanotubes designed in our laboratory used the cyclic peptide cyclo-[(Gln-D-Ala-Glu-D-Ala)2], which contains two glutamic acid residues. This allowed us to have a pH triggered control over self assembly[7a]. At neutral or basic pH the glutamic acid residues are deprotonated and negatively charged, which causes the cyclic peptide to be soluble in aqueous solution and furthermore creates a repulsive electrostatic force between two cyclic peptides if they approach one another. When the pH is dropped below 5 the glutamic acid residues become protonated thereby losing their negative charge. This decreases the peptide's solubility and removes the repulsive forces allowing the cyclic peptides to come together in a highly concerted fashion to form nanotubes.

When the pH of such a solution is dropped, a milky white suspension forms. When observed at a magnification of 100 or more the suspension is seen to be made up of millions of microcrystalline rod shaped objects which are dozens of microns long and a few microns wide. When observed by cryo-electron microscopy at magnifications above 35,000 one can clearly see striations along the long axis of these microcrystals which are spaced by about 15 angstroms and correspond to the diameter of the cyclic peptide. We believe these striation are the individual nanotubes lying parallel to one another in a microcrystalline bundle.

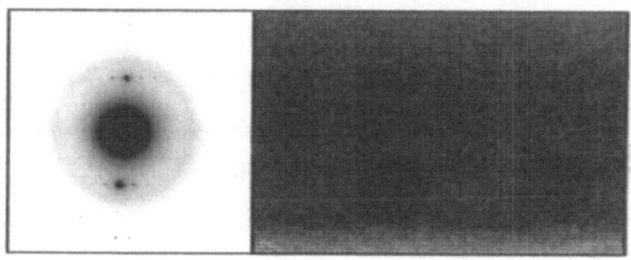

Figure 2) Electron diffraction pattern (left) and high magnification electron micrograph (right) of cyclo-[(Gln-D-Leu)4]. Structure factors in the diffraction pattern give the unit cell data (angles and size) which are used for modeling. Striations in the electron micrograph are spaced by 18Å which corresponds to the diameter of the cyclic peptide building block.

Although the pH triggered self-assembly is an elegant way to control nanotube formation, it is also a liability because those nanotubes are not stable in neutral or basic conditions and are, in general, fragile. In order to address this problem we prepared a series of cyclic peptides in which the glutamic acid residues were replaced with glutamine and the nonpolar residue was replaced with a series of hydrophobic residues including alanine, valine, leucine and phenylalanine[7d]. These nanotubes were self-assembled by dissolving the peptide in trifluoroacetic acid (TFA), the only solvent in which these cyclic peptides are found to be universally soluble, and slowly adding water to initiate self-assembly. These nanotubes were found to be very robust and stable to a wide range of pH and solvents as well as to physical stress. Similar striated microcrystals were visible by cryo-transmission electron microscopy (TEM).

The cryo-TEM images together with electron diffraction patterns, which provide us the unit cell dimensions and angles, allow us to prepare models of these nanotubes. Several reasonable models of each nanotube were prepared with the program XtalView[10] within the unit cell provided from the electron diffraction data. Simulated diffraction spectra were calculated from these models and were compared to the actual diffraction data. The models with the best match were subjected to examination for bad contacts using Procheck[11].

Figure 3) Model of *cyclo*-[(Gln-*D*-Leu)4]. The inner diameter of the cyclic peptide measures 7.5Å. The hydrophobic groups (leucine) face one another making favorable van Der Waals contacts while the hydrophilic groups (glutamine) are able to fulfill their hydrogen bonding potential.

The models of *cyclo*-[(Gln-*D*-Ala-Glu-*D*-Ala)2], *cyclo*-[(Gln-*D*-Ala)4], *cyclo*-[(Gln-*D*-Leu)4], *cyclo*-[(Gln-*D*-Phe)4-] were found to be quite similar. In all cased the central channel remains open with a diameter of about 7.5 angstroms. The hydrophobic groups faced one another forming favorable hydrophobic contacts while the glutamine or glutamic acid residues faced toward one another in groups of four where they can satisfy their hydrogen bonding potential.

Using techniques similar to those previously mentioned, a twelve amino acid cyclic peptide, *cyclo*-[(Gln-*D*-Ala-Glu-*D*-Ala)3], was synthesized and self assembled into nanotubes[7b]. In contrast to the eight amino acid cyclic peptide based nanotubes, this nanotube forms an hexagonal lattice, but overall the packing is similar to the eight membered cyclic peptides. These examples illustrate the ease by which tube diameter and outer chemistries of cyclic peptide-based nanotubes can be controlled.

BASIC STRUCTURAL REPEAT

The basic repeating unit of peptide nanotubes is a dimer of cyclic peptides which associate in an antiparrallel fashion. In order to more closely examine the structure of this unit and the strength of its association a cyclic peptide in which the D-amino acids had been N-methylated was prepared[12]. This N-methylation prevents the propagation of hydrogen bonding from extending beyond a dimer and prevents parallel hydrogen bonding altogether. One example of this type of peptide dimer, *cyclo*-[(D-MeNAla-Phe)4-], was crystallized and its X-ray crystal structure was used to confirm the models of our other nanotubes as it illustrated definitively that these cyclic peptides do in fact adopt a flat, open conformation. Further studies with this peptide were used to demonstrate that these cyclic peptides do in fact prefer an antiparallel association. *Cyclo*-[(D-MeNAla-Phe)4-] along with its enantiomer were mixed and variable temperature NMR was used to determine the association constants of self association (anti-parallel) and enantiomeric association (parallel). This study showed a clear preference for anti-parallel over parallel β-sheet association[13].

Figure 4) p and q enantiomers of the cyclic peptide were prepared and association constants between p-p, q-q and p-q were calculated by variable temperature NMR experiments.

MEMBRANE CHANNELS

Complete control over outer surface property of nanotubes can be had by altering the amino acids that make up the cyclic peptides. Thus nanotubes were made which are hydrophobic and selectively partition into a lipid bilayer. In this non-polar environment, hydrogen bonding is highly favored and helps to drive self assembly within the membrane. Once formed, this transmembrane channel functioned as an effective transporter of ions and

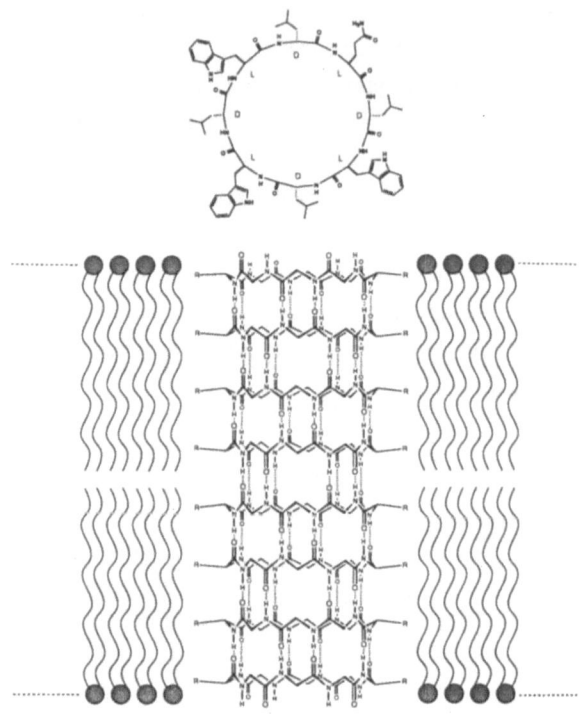

Figure 5) Hypothetical structure of trans-membrane nanotube ion-channel.

small molecules. This function was illustrated in two studies. First we looked at the pH gradien driven transport of protons across the membrane of unilaminar liposomes. This was monitored by incorporating 5(6)-carboxyfluorescein within the liposome whose florescence changes with pH. The exterior of the liposome had a pH of 5.5 while the interior a pH of 6.5. Open channels should rapidly equilibrate the pH between these two compartments. The cyclic peptide used, cyclo-[Gln-D-Trp-(Leu-D-Trp)3], was highly hydrophobic having only a single polar residue (glutamine) in order to simplify its synthesis. For comparison the natural ion channel gramacidin A and amphotericin B were also tested. The results showed that cyclic peptide based nanotube was indeed an effective proton channel as it equilibrated the pH between the interior and exterior of the liposome as fast or faster then the natural ion channels tested.

As a control the cyclic peptide cyclo-[(Gln-D-Leu)4], which can form nanotubes but is not hydrophobic enough to partition within the lipid bilayer and cyclo-[(D-MeNAla-Phe)4], which is can partition into the lipid bilayer but cannot form tubes, were also tested and found not to have channel activity.

The best way to test for ion channel activity, however, is by the patch-clamp method which looks at single channel conductance measurements. When the nanotube made from *cyclo*-[Gln-*D*-Trp-(Leu-*D*-Trp)3] was tested in KCl and NaCl solutions, conductances on the

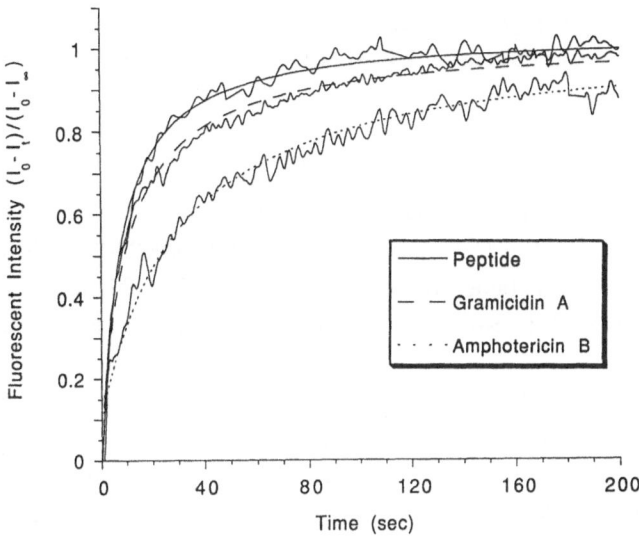

Figure 6) Plot of fluorescent intensity as the pH between the interior and exterior of liposomes is equilibrated by peptide nanotubes (solid line), gramicidin A (dashed line) and amphotericin B (dotted line)

order of 60 picoseimens were recorded which correspond to the transport of approximately 20 million ions per second[14].

Small molecule transport, specifically glucose, across a lipid bilayer was monitored using the following enzymatic assay. Glucose containing liposomes were placed in a solution containing the enzymes hexokinase and glucose-6-dehydrogenase in addition to ATP and NADP. As glucose is released, it is phosphorylated by hexokinase and then oxidized by glucose-6-dehydrogenase. In order to oxidize glucose-6-phosphate, NADP is reduced to NADPH and it is the UV absorbance of NADPH that is monitored as an indication of glucose transport. This was attempted with the 8 amino acid cyclic peptide, however no transport was observed which was expected based on its relatively small inner pore diameter. A 10 amino acid cyclic peptide, *cyclo*-[Gln-*D*-Trp-(Leu-*D*-Trp)4], was prepared and it was successful in transporting glucose thanks to it larger, 10 angstrom pore[15]. These results demonstrate the ability of peptide nanotubes to transport small molecules in a size selective fashion across lipid bilayers.

CONCLUSION

We have shown structural characterization and functional studies on a variety of cyclic peptide-based nanotubes in which we have demonstrated control over nanotube diameter and outer surface chemistries. The versatility and demonstrated functionality of these nanotubes, coupled with their ease of synthesis makes them potential candidates for such uses as antibiotics and drug delivery vehicles and in the design of nanowires, novel catalyst supports and inclusion and separation technologies.

1) Harada A, Li J, Kamachi M (1993) Nature 364:516

2) Monnier A, Schüth F, Huo Q, Kumar D, Margolese D, Maxwell RS, Stucky GD, Krishnamury M, Petroff P, Firouzi A, Janicke M, Chmelka BF (1993) Science 261:1299

3) Meier WM, Olson DH (1988) Atlas of Zeolite Structure Types 2nd edn, Butterworths, London

4) Iijima S, Nature (1991) 354:56

5) De Santis P, Morosetti S, Rizzo R (1974) Macromolecules 7:52

6a) Tomasic L, Lorenzi GP, (1978) Helv Chim Acta 70:1012 b) Pavone V, Benedetti E, Di Blasio B, Lombardi A, Pedone C, Tomasich L, Lorenzi GP (1989) Biopolymers 28:215

7a) Ghadiri MR, Granja JR, Milligan RA, McRee DE, Khazanovich N (1993) Nature 366:324 b) Khazanovich N, Granja JR, McRee DE, Milligan RA, Ghadiri MR, (1994) J Am Chem Soc 116:6011 c) Ghadiri MR (1995) Adv Mater 7:675 d) Hartgerink JD, Granja JR, Milligan RA, Ghadiri MR (1995) J Am Chem Soc, submitted

8) Schnolzer M, Alewood P, Jones A, Alewood D, Kent SBH, (1992) Int J Peptide Protein Res 40:180

9) Rovero P, Quartara L, Fabbri G (1991) Tet Lett 32:2639

10) McRee DE (1992) J Molec Graphics 10:44

11) Laskowski RA, MacArthur MW, Moss DS, Thorton JM, (1993) J Appl Cryst 26:283

12) Ghadiri MR, Kobayashi K, Granja JR, Chadha RK, McRee DE (1995) Angew Chem Int Ed Engl 34:93

13) Kobayashi K, Granja JR, Ghadiri MR (1995) Angew Chem Int Ed Engl 34:95

14) Ghadiri MR, Granja JR, Buehler LK (1994) Nature 369:301

15) Granja JR, Ghadiri MR (1994) J Am Chem Soc 116:10785

Development of a Rapid and Facile Method for Protein Synthesis Using Partially Protected Peptide Thioesters as Building Blocks

Saburo Aimoto and Hironobu Hojo

Institute for Protein Research, Osaka University,
3-2 Yamadaoka, Suita, Osaka 565 Japan

INTRODUCTION

The development of a new synthetic strategy is essential to realize routine preparation of proteins by chemical means. Its success will also open a way to create proteins with novel characteristics and functions, which will not be attained by a recombinant DNA technology. Using a solution method and a solid-phase method (1), a tremendous number of complex peptides have been synthesized. Although both methods have been successfully applied to the preparation of small peptides, most protein preparation has not been performed by chemical means, but by recombinant DNA technology, because of intrinsic difficulties involved in both chemical methods.

In the solution method, the product has to be isolated and confirmed after every coupling reaction. Thus, this method of synthesis requires numbers of experienced chemists. Nevertheless, synthesis is not always successful because of solubility and/or reactivity problems of protected intermediate peptides. Several groups have synthesized proteins by solution method, such as RNase A (2) and parathyroid hormone.(3) However, those procedures cannot form the general basis of protein synthesis because of the reasons mentioned above.

On the other hand, the solid-phase method realized simple and rapid preparation of peptides. Reagents are introduced to the reaction vessel containing a peptide on an insoluble resin and mixed with it for a given time. After that, the resin is washed with appropriate solvents to remove excess reagents. Basic chemical processes have been intensively studied and sophisticated protocols for chain elongation cycles have been developed for machine assisted synthesis. Along with the development of related technologies such as reversed-phase high performance liquid chromatography (RPHPLC) and mass spectrometry, the solid-phase method became the major means of peptide synthesis. The solid-phase method, however, also presents problems in the protein synthesis. For instance, inter-chain aggregation and secondary-structure formation occur more or less with the increase in peptide-chain length. This causes imperfect deprotection of an α-amino protecting group as well as incomplete incorporation of an amino-acid derivative. Mainly because of these intrinsic difficulties, the synthesis of a highly pure peptide with more than 50 amino acid residues is still difficult.

M. Kamachi · A. Nakamura (Eds)
New Macromolecular Architecture and Functions
Proceedings of the OUMS '95 Toyonaka, Osaka, Japan, 2-5 June, 1995
© Springer-Verlag Berlin Heidelberg 1996

Thus, a new strategy has to be developed for the rapid synthesis of highly pure protein. Several groups are developing new methods, in which protected peptide segments are prepared by the solid-phase method and then used for segment condensation in solutions.(4-7) In their strategies, the protected peptide segments have to be designed taking the later condensation method in solution into account.

Among these methods, the thiocarboxyl segment condensation strategy (4) has an attractive feature. A peptide with thiocarboxylic acid at the carboxyl terminus was prepared by the solid-phase method and the segment was condensed with an amino component in the presence of silver ions and *N*-hydroxysuccinimide (HONSu). In the method, no protecting group is required for the side chain carboxyl group, for the thiocarboxyl group is selectively activated by silver ions. Using this strategy, several proteins were synthesized.(8-12) However, the method also has a limitation. Thiocarboxylic acid is easily decomposed by oxidation or hydrolysis. Furthermore, the amino protective reagent, such as *N*-(*t*-butoxycarbonyloxy)succinimide (Boc-ONSu), could not be used to protect side-chain amino groups because of the high nucleophilicity of the thiol moiety of the thiocarboxyl group. Only citraconic anhydride could be used to introduce a citraconyl group into the side-chain amino groups. Citraconyl is, however, unstable even under mild acidic conditions and it does not increase the solubility of the protected peptide even in a polar organic solvent such as dimethyl sulfoxide (DMSO). Furthermore, no route could be developed to prepare cysteine-containing proteins based upon the thiocarboxyl segment condensation procedure.

To overcome the problems inherent in the Blake's method, We designed a new method for protein synthesis, in which the *S*-alkyl thioester of a partially protected peptide segment is used as a building block. In this paper, we describe the development of the method for protein synthesis according this idea using the *S*-alkyl thioester of partially protected peptide segments.

RESULTS AND DISCUSSION

The thioester group in a peptide segment was expected to be activated by silver ions. If so, the terminal thioester group will be selectively activated. Furthermore the thioester group is not as nucleophilic as the thiocarboxyl group. Hence various kinds of protecting groups can be introduced to the side-chain amino groups using protective reagents such as Boc-ONSu. If the *S*-alkyl thioester of a peptide segment could be synthesized by a solid-phase method in a high yield, a very promising novel method for protein synthesis would be developed. To establish such a method, we synthesized proteins searching conditions to realize it.

Synthesis of *c*-Myb Protein(142-193)-NH$_2$ (13)

The thioester strategy consists of two steps; the preparation of a partially protected peptide thioester via a solid-phase method and segment condensation in the presence of silver ions in a homogeneous solution. Thioester is a kind of active esters. Hence, its stability has to be confirmed throughout the preparation processes of partially protected peptide thioesters. Furthermore, the thioester moiety in a peptide segment has to be converted to the corresponding active ester efficiently in the presence of silver ions. All the processes involved in the thioester method were examined by synthesizing the DNA-binding domain of c-Myb protein (14), namely c-Myb protein(142-193)-NH$_2$. The amino acid sequence is shown in Fig. 1. Two partially protected peptide segments corresponding to the sequences of 142-163 and 164-193 were prepared and were coupled to form the sequence 142-193.

```
142
Val-Lys-Lys-Thr-Ser-Trp-Thr-Glu-Glu-Glu-Asp-Arg-Ile-
                                      ↓
Ile-Tyr-Gln-Ala-His-Lys-Arg-Leu-Gly-Asn-Arg-Trp-Ala-

Glu-Ile-Ala-Lys-Leu-Leu-Pro-Gly-Arg-Thr-Asp-Asn-Ala-
                                                  193
Ile-Lys-Asn-His-Trp-Asn-Ser-Thr-Met-Arg-Arg-Lys-Val
```

Fig. 1. Primary sequence of c-Myb protein(142-193) deduced from a cDNA clone of murine c-*myb* mRNA.(17) Arrow indicates the site of segment coupling

Preparation of Boc-[Lys(Boc)[143,144,160]]-c-Myb Protein(142-163)-SCH$_2$CH$_2$CONH$_2$ (1): The preparation of Peptide **1** is summarized in Fig. 2. Starting from Boc-Gly-SCH$_2$CH$_2$CONH-resin, a protected c-Myb protein(142-163)-SCH$_2$CH$_2$CONH-resin was prepared on a peptide synthesizer model 430A (Applied Biosystems Inc.) according to the protocol of the system software version 1.40 NMP/HOBt t-Boc. This resin was treated by anhydrous HF(15) in the presence of p-cresol and 1,4-butanedithiol.(16) Judging from the measurement of mass number of peptides separated by RPHPLC, the thioester was stable under HF-treatment. The crude peptide was purified by RPHPLC to yield highly pure peptide **5**. The yield was 17% based on the glycine in the starting resin. The yield of peptide **5** was about one-half that of peptide amide **6** (30%) synthesized on 4-methylbenzhydrylamine (MBHA) resin as described later. This difference could be due to partial decomposition of the linker moiety during the chain elongation reaction. The thioester in peptide **5** proved to be quite stable during purification on RPHPLC and during prolonged storage at 4 °C. The introduction reaction of Boc groups to peptide **5** proceeded almost quantitatively and did not accompany any side reactions when Boc-ONSu was used as the protective reagent. The yield of peptide **1** was practically quantitative.

Preparation of [Lys(Boc)[171,182,192]]-c-Myb Protein(164-193)-NH$_2$ (2): A peptide chain of c-Myb protein(164-193) was assembled on MBHA resin according to the same protocol described

Boc-Gly-SCH₂CH₂CONH-resin

> ABI 430A Peptide synthesizer
> System software version 1.40 NMP/HOBt *t*-Boc.
> End capping by acetic anhydride.

Boc-Val-Lys(Cl-Z)-Lys(Cl-Z)-Thr(Bzl)-Ser(Bzl)-Trp(For)-Thr(Bzl)-Glu(OBzl)-Glu(OBzl)-
Glu(OBzl)-Asp(OcHex)-Arg(Tos)-Ile-Ile-Tyr(Br-Z)-Gln-Ala-His(Bom)-Lys(Cl-Z)-Arg(Tos)-
Leu-Gly-SCH₂CH₂CONH-resin

> 1) HF treatment, 2) RPHPLC

Val-Lys-Lys-Thr-Ser-Trp-Thr-Glu-Glu-Glu-Asp-Arg-Ile-Ile-Tyr-Gln-Ala-His-Lys-Arg-Leu-Gly-
SCH₂CH₂CONH₂ (**5**)

> Boc-ONSu

Boc-Val-Lys(Boc)-Lys(Boc)-Thr-Ser-Trp-Thr-Glu-Glu-Glu-Asp-Arg-Ile-Ile-Tyr-Gln-Ala-His-
Lys(Boc)-Arg-Leu-Gly-SCH₂CH₂CONH₂ (**1**)

Fig. 2.   Synthetic scheme of Boc-[Lys(Boc)[143,144,160]]-*c*-Myb Protein(142-163)-
SCH₂CH₂CONH₂ (**1**)

for the synthesis of peptide **1**. After the completion of chain assembly, a 2,2,2-
trichloroethoxycarbonyl (Troc) group was introduced to protect the terminal amino group of
peptide on a resin using *N*-(2,2,2-trichloroethoxycarbonyloxy)succinimide (Troc-ONSu).   After
HF treatment of this protected peptide resin followed by HPLC purification, Troc-Asn-Arg-Trp-
Ala-Glu-Ile-Ala-Lys-Leu-Leu-Pro-Gly-Arg-Thr-Asp-Asn-Ala-Ile-Lys-Asn-His-Trp-Asn-Ser-
Thr-Met-Arg-Arg-Lys-Val amide (**6**) was obtained in a yield of 30% based on the amino group
in the starting MBHA resin.   To peptide **6,** Boc groups were introduced to block the side-chain
amino groups.   The peptide thus obtained was treated by zinc dust in aqueous acetic acid to give
peptide amide **2** in an overall yield of 16%.

Preparation of c-Myb Protein(142-193)-NH₂ (4):   As shown in Fig. 3 the thioester group in
peptide **1** was converted to the corresponding active ester in the presence of 4-nitrophenol
(HONp) and AgNO₃ in DMSO.   To the DMSO solution, peptide amide **2** and 4-
methylmorpholine (NMM) were added.   The progress of the coupling reaction was monitored by
RPHPLC.   After 3 days, the condensation reaction was almost completed.   A product was treated
with trifluoroacetic acid (TFA).   After removal of TFA, a product was isolated by RPHPLC at
the yield of 50%. This yield is quite high, if the nonspecific adsorption of peptide amide **4** during
the isolation process, which usually occurs in peptide synthesis, is taken into account.   The
methionine was not damaged by silver ions under the reaction conditions employed.   No other
serious side reactions were observed during segment coupling under this minimum protection
strategy either.

Boc-[Lys(Boc)^{143,144,160}]-c-Myb protein(142-163)-SCH$_2$CH$_2$CONH$_2$ (**1**)

    │   AgNO$_3$ + HONp + NMM
    ▼

Boc-[Lys(Boc)^{143,144,160}]-c-Myb protein(142-163)-ONp

    │   [Lys(Boc)^{171,182,192}]-c-Myb protein(164-193)-NH$_2$ (**2**)
    │   + NMM

Boc-[Lys(Boc)^{143,144,160,171,182,192}]-c-Myb protein(142-193)-NH$_2$ (**3**)

    │   TFA containing HS(CH$_2$)$_4$SH (5% v/v)
    ▼

c-Myb protein(142-193)-NH$_2$ (**4**)

Fig. 3. Synthetic route leading to c-Myb Protein(142-193)-NH$_2$ by segment coupling.

Use of partially protected peptide thioester as a building block: Partially protected peptide segments **1** and **2** were successfully prepared and well characterized by amino acid analysis and fast atom bombardment (FAB) mass spectrometry. The thioester group of peptide **1** was efficiently and selectively activated by silver ions and converted to the corresponding active ester. As a result, protection of the side-chain carboxyl groups was unnecessary. Only the functional groups that required protection were amino group in this synthesis. Protecting groups such as the Boc group increases the solubility of a partially protected peptide segment in DMSO used for the subsequent segment coupling step. Only one protecting group existed on peptide **5** or **6** when purified on RPHPLC. Consequently, the purification was much more effective, compared with that in the BPTI synthesis.(5) Therefore long peptide segments such as **5, 6** were easily prepared. The yield of segment coupling was also satisfactory. Thus, we can conclude that the partially protected peptide thioester is a promising building block for polypeptide synthesis .

Development of a Thioester Linker with Enhanced Stability (17)

To make the thioester method more effective for the synthesis of proteins, improvements are required of several points. One of the major drawbacks is the low yield of peptide thioester due to the instability of the linker containing a thioester moiety on a solid support.

Development of a linker containing an S-alkyl-thioester-moiety with enhanced stability: Chemical characteristics of linkers containing *S-n*-alkyl, *S-s*-alkyl or *S-t*-alkyl thioester were compared in order to estimate the stability of the thioester-containing linker during peptide chain elongation cycles. Each crude product was analyzed by RPHPLC after an HF treatment. The results are summarized in Table 1. A pentapeptide with an *S-t*-alkyl thioester group was obtained in 35% yield, whereas the others were obtained in lower yields. The difference in the yield seems to be amplified by the 2,5-piperazinedione formation of a Pro-Gly sequence at the carboxyl terminal of the peptide. These results suggest that the linker containing the *S-t*-alkyl thioester has favorable characteristics for the preparation of the peptide thioester in regard to the yield and avoiding 2,5-piperazinedione formation.

Table 1. Analysis of the crude products obtained using three thioester moieties

	Yield/%	
-SR	Trp-Lys-His-Pro-Gly-SR	Trp-Lys-His-SR
A) -SCH$_2$CH$_2$CONH$_2$	16	8
B) -SCH(CH$_3$)CH$_2$CONH$_2$	30	5
C) -SC(CH$_3$)$_2$CH$_2$CONH$_2$	35	0

Stability test of Boc-Gly-SC(CH$_3$)$_2$CH$_2$CONH-Resin: In order to analyze the factors responsible for the yield of the peptide *S-t*-alkyl thioester, Boc-Gly-SC(CH$_3$)$_2$CH$_2$CONH-resin was treated under the conditions used for peptide chain-elongation cycles. When Boc-Gly-SC(CH$_3$)$_2$CH$_2$-CONH-resin was treated with 50% TFA in DCM (v/v), Gly-SC(CH$_3$)$_2$CH$_2$CONH$_2$ was liberated at the rate of 2 to 3% per one amino acid elongation cycle. Under the same conditions, Gly-NH$_2$ was removed from the Boc-Gly-NH-resin at a rate of 0.1% per cycle. On the other hand, the peptide thioester resin was rather stable in the presence of 5% DIEA in DMF (v/v). To estimate the effect of the sulfur atom upon the stability of the thioester-containing linker on the MBHA resin, Nle or β-Ala was inserted as a spacer between the thioester moiety and the MBHA resin. Using Boc-Gly-SC(CH$_3$)$_2$CH$_2$CO-Nle-(or β-Ala-)NH-resin, Troc-HBs(16-39) thioesters were synthesized. The yields were compared with those of peptide thioesters without spacer groups. The presence of the spacer groups, β-Ala or Nle, enhanced the stability of the linkers on the MBHA resin as shown in Table 2. The yields of the peptide

Table 2. Syntheses of the peptide thioester of Troc-HBs(16-39)
using 4 resins with different thioester-containing linkers

Troc-Leu-Ser-Lys-Lys-Asp-Ala-Thr-Lys-Ala-Val-Asp-Ala-Val-Phe-Asp-Ser-Ile-Thr-Glu-Ala-Leu-Arg-Lys-Gly-SR	
-SR	Yield / %
-SCH$_2$CH$_2$CONH$_2$	15
-SC(CH$_3$)$_2$CH$_2$CONH$_2$	15
-SC(CH$_3$)$_2$CH$_2$CONH-β-Ala-NH$_2$	26
-SC(CH$_3$)$_2$CH$_2$CO-Nle-NH$_2$	28

thioesters with these spacers were almost equal to that of a peptide amide prepared on an MBHA resin without a thioester moiety.

Synthesis of Barnase Site-Specifically Labeled with Two ^{13}C Atoms (18)

To estimate the usefulness of the thioester method, the method was applied to the synthesis of barnase, a protein comprising 110 amino acids with RNase activity (Fig. 4).(19) In this study, barnase site-specifically labeled with two ^{13}C atoms was synthesized for future structural studies.

```
 1                                          10
Ala-Gln-Val-Ile-Asn-Thr-Phe-Asp-Gly-Val-Ala-Asp-Tyr-Leu-Gln-
              20                                          30
Thr-Tyr-His-Lys-Leu-Pro-Asp-Asn-Tyr-Ile-Thr-Lys-Ser-Glu-Ala-
                  ↓                       40
Gln-Ala-Leu-Gly-Trp-Val-Ala-Ser-Lys-Gly-Asn-Leu-Ala-Asp-Val-
              50      *   ↓                               60
Ala-Pro-Gly-Lys-Ser-Ile-Gly-Gly-Asp-Ile-Phe-Ser-Asn-Arg-Glu-
                                          70              *
Gly-Lys-Leu-Pro-Gly-Lys-Ser-Gly-Arg-Thr-Trp-Arg-Glu-Ala-Asp-
              80      ↓                                   90
Ile-Asn-Tyr-Thr-Ser-Gly-Phe-Arg-Asn-Ser-Asp-Arg-Ile-Leu-Tyr-
                                         100
Ser-Ser-Asp-Trp-Leu-Ile-Tyr-Lys-Thr-Thr-Asp-His-Tyr-Gln-Thr-
                     110
Phe-Thr-Lys-Ile-Arg
```

Fig. 4. The amino acid sequence of barnase. The arrows indicate the sites of segment coupling; The asterisks indicate amino acids labeled with ^{13}C; (2-^{13}C)Gly^{52} and (1-^{13}C)Ala^{74}

Preparation of peptide segments: For synthetic purposes, the barnase sequence was divided into four peptide segments, as shown in Fig. 4. A partially protected peptide thioester was prepared as follows: To an MBHA resin, Boc-Nle and Boc-Gly-SC(CH$_3$)$_2$CH$_2$COOH were successively introduced using DCC in the presence of HOBt to obtain Boc-Gly-SC(CH$_3$)$_2$CH$_2$CO-Nle-NH-resin. On this resin, Boc-amino acids were successively condensed. After peptide chain assembly, the terminal amino group was protected with an 4-pyridylmethoxycarbonyl (*i*Noc) group. The protected peptide resin was treated with anhydrous HF to give a crude *i*Noc-peptide thioester, which was purified by RPHPLC. Boc groups were introduced to the side-chain amino groups of an HPLC-purified peptide thioester in order to realize selective removal of the amino protecting groups after segment condensation. The partially protected peptide segments were prepared in good yields. All the partially protected peptide segments used for barnase synthesis are listed in Table 3. The yields of the peptide segments were calculated based upon the amino groups in the MBHA resin. The linker containing Nle and *S-t*-alkyl thioester moieties gave satisfactory yields.

Table 3. Partially protected peptide segments prepared for segment coupling

Peptide segment	Yield/%[a]
Boc-[Lys(Boc)[19,27]]-Barnase(1-34)-SC(CH₃)₂CH₂CO-Nle-NH₂ (7)	12

$$Boc\text{-}[Lys(Boc)^{19,27}]\text{-}Barnase(1\text{-}34)\text{-}SC(CH_3)_2CH_2CO\text{-}Nle\text{-}NH_2 \quad (7)$$

Boc-Ala-Gln-Val-Ile-Asn-Thr-Phe-Asp-Gly-Val-Ala-Asp-Tyr-Leu-Gln-Thr-Tyr-His-
Lys(Boc)-Leu-Pro-Asp-Asn-Tyr-Ile-Thr-Lys(Boc)-Ser-Glu-Ala-Gln-Ala-Leu-Gly-
SC(CH₃)₂CH₂CO-Nle-NH₂

iNoc-[Lys(Boc)^{39,49}, (2-13C)Gly^{52}]-Barnase(35-52)-SC(CH₃)₂CH₂CO-Nle-NH₂ (8)　32
　iNoc-Trp-Val-Ala-Ser-Lys(Boc)-Gly-Asn-Leu-Ala-Asp-Val-Ala-Pro-Gly-Lys(Boc)-
　Ser-Ile-Gly*-SC(CH₃)₂CH₂CO-Nle-NH₂

iNoc-[Lys(Boc)^{62,66}, (1-13C)Ala^{74}]-Barnase(53-81)-SC(CH₃)₂CH₂CO-Nle-NH₂ (9)　12
　iNoc-Gly-Asp-Ile-Phe-Ser-Asn-Arg-Glu-Gly-Lys(Boc)-Leu-Pro-Gly-Lys(Boc)-Ser-Gly-
　Arg-Thr-Trp-Arg-Glu-Ala*-Asp-Ile-Asn-Tyr-Thr-Ser-Gly-SC(CH₃)₂CH₂CO-Nle-NH₂

[Lys(Boc)^{98,108}]-Barnase(82-110)　(10)　12
　Phe-Arg-Asn-Ser-Asp-Arg-Ile-Leu-Tyr-Ser-Ser-Asp-Trp-Leu-Ile-Tyr-Lys(Boc)-Thr-
　Thr-Asp-His-Tyr-Gln-Thr-Phe-Thr-Lys(Boc)-Ile-Arg

a) The yield was calculated based on the amino groups in the MBHA resin.

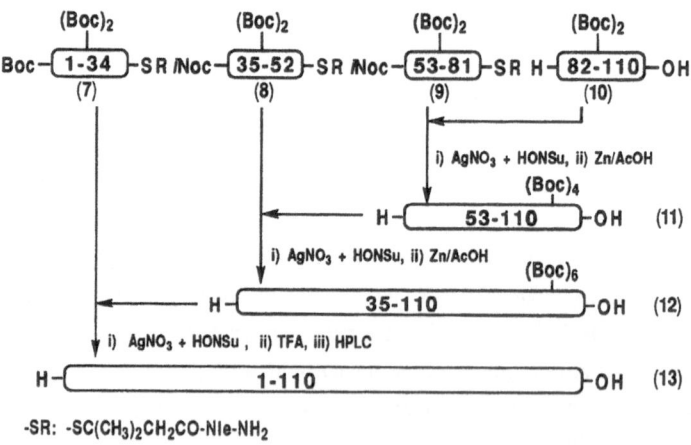

Fig. 5. Synthetic route of barnase(1-110)

Synthesis of barnase by segment coupling: Segment condensation was performed according to the scheme shown in Fig. 5. The typical coupling conditions were as follows. Peptides **9** (17 mmol) and **10** (13 mmol) were dissolved in DMSO (2.3 ml). HONSu (260 mmol), AgNO₃ (77 mmol), and NMM (82 mmol) were then added in succession. The solution was stirred overnight at room temperature. The peptide was precipitated with distilled water and washed. *i*Noc group was removed by zinc dust treatment. The supernatant was dialyzed against distilled water and freeze-dried to give a mixture containing peptide 11. According to a similar procedure, peptides 8 and 11, then peptides 7 and 12, were successively condensed. The condensation reactions were monitored by RPHPLC using a C4 column. The segment coupling of peptide 11 and 8 and that of peptide 12 and 7 were almost complete within 6 hr. During peptide chain elongation by segment coupling, no RPHPLC purification was performed. After segment condensation of peptides 7 and 12, a product was precipitated by adding distilled water and freeze-dried to give a powder that was treated with TFA. The crude product was purified on PROTEIN-RP, followed by Pharmacia HiLoad S-Sepharose HP to give the final product, barnase(1-110) (13) in 11% yield, based upon peptide 10. The RNase activity of the synthetic barnase was determined by the method described by Rushizky et al..(20) The synthetic and native barnases had practically same activities.

Evaluation of the Thioester Method

In the barnase synthesis, all the segment condensation reactions were almost completed within 6 hr, even those between peptides 7 and 12 of 34 and 76 amino acid residues, respectively. This fact suggests that the peptide segment with minimum protecting groups was well solvated by DMSO, kept good flexibility around the reaction sites and, hence, retained high reactivity. Thus, the thioester method, which uses a minimal protection strategy, is suitable for protein synthesis, not only because of the ease of segment preparation, but also because of the high reactivity during segment condensation.

Taking the results obtained by barnase synthesis into account, we searched an easily removable protecting group instead of Troc or *i*Noc for the terminal amino group. We examined an 9-fluorenylmethoxycarbonyl (Fmoc) group for the terminal amino protection. This protecting group was easily removed by piperidine treatment and was stable under the segment coupling conditions. Thus, Fmoc was superior to Troc or *i*Noc groups as a terminal amino protecting group, because we can avoid the use of zinc-dust.

We also developed a new strategy, with which cysteine-containing proteins can be synthesized. Using the strategy, we synthesized the barnase-like domain of DNA-directed RNA-polymerase II of *Saccharomyces cerevisiae*, consisting of 112 amino-acid residues (RPSc(299-410)).(21) Partially protected cysteine-containing peptide thioesters were successfully prepared by a solid-phase method using Npys-amino acids.(22) These segments could be condensed in the presence of silver ions without any loss of the S-4-methylbenzyl and S-2,4,6-trimethylbenzyl groups on mercapto groups of cysteine residues and we obtained the desired product, which was isolated as a distinct peak by RPHPLC. (23)

Thus, the thioester method has proven useful for the facile preparation of proteins with or without cysteine residue and is expected to open a way to create proteins with novel characteristics and functions. Furthermore, the thioester method will provide a very promising basis for the synthesis of conjugated proteins such as phosphoprotein and glycoprotein because of its minimum protection strategy.

References

1 Merrifield RB (1963) J. Am. Chem. Soc 85:2149

2 Yajima H, Fujii N (1980) J. Chem. Soc., Chem. Commun 1980:115

3 Kimura T, Takai M, Yoshizawa K, Sakakibara S (1983) Biochem. Biophys. Res. Commun 114:493

4 Blake J, Li CH (1981) Proc. Natl. Acad. Sci. USA 78:4055

5 Aimoto S, Mizoguchi N, Hojo H, Yoshimura S (1989) Bull. Chem. Soc. Jpn 62: 524

6 Kaiser ET, Mihara H, Laforet GA, Kelly JW, Walters L, Findeis MA, Sasaki T(1989) Science 243:187

7 Sabatier JM, Tessier-Rochat M, Granier C, Rietschoten JV, Pedrodo E, Grandas A, Albericio F, Giralt E (1987) Tetrahedron 43:5973

8 Blake J, Li CH (1983) Proc. Natl. Acad. Sci. USA 80: 1556

9 Blake J, Westphal M, Li, CH (1984) Int. J. Peptide Protein Res 24:498

10 Blake J (1986) Int. J. Peptide Protein Res 27:191

11 Blake J, Yamashiro D, Ramasharma K, Li, CH (1986) Int. J. Peptide Protein Res 28: 468

12 Yamashiro D, Li CH (1988) Int. J. Peptide Protein Res 31:322

13 Hojo H, Aimoto S (1991) Bull. Chem. Soc. Jpn 64 :111

14 Gonda JT, Gough NM, Dunn AR, de Blaquiere J (1985) EMBO J 4:2003

15 Sakakibara S, Shimonishi Y, Kishida Y, Okada M, Sugihara H (1967) Bull. Chem. Soc. Jpn 40:2164

16 Aimoto S, Shimonishi Y (1976) Bull. Chem. Soc. Jpn 49 :317

17 Hojo H, Kwon YD, Kakuta Y, Tsuda S, Tanaka I, Hikichi K, Aimoto S (1993) Bull. Chem. Soc. Jpn 66:2700

18 Hojo H, Aimoto S (1993) Bull. Chem. Soc. Jpn 66:3004

19 Paddon CJ, Hartley RW (1986) Gene 40 :231

20 Rushizky GW, Greco AE, Hartley RW, Sober HA (1963) Biochemistry 2:787

21 Shirai T, Go M (1991) Proc. Natl. Acad. Sci. USA 88:9056

22 Matsueda M, Walter R (1980) Int. J. Peptide Protein Res 16:392

23 Hojo H, Yoshimura S, Go M, Aimoto S (1995) Bull. Chem. Soc. Jpn 68:330

Molecular Weight Dependent Antimicrobial Activity by Chitosan

Seiichi Tokura, Keisuke Ueno, Satoshi Miyazaki and Norio Nishi
Graduate School of Environmental Earth Science, Hokkaido University,
Sapporo 060, Japan

Chitosan oligomers of average molecular weight 9300(P9300) and 2200(P2200) were prepared by nitrous acid degradation followed by the reduction of 2,5-anhydromannose terminal by sodium borohydrate. Although P9300 provided the growth inhibition of *Escherichia coli*, P2200 was not growth inhibitor but growth accelerator. The stacking of P9300 to cell wall of *E. coli* was confirmed by the use of FITC(Fluorescein isothiocyanate) labeled chitosan oligomer using the Confocal Laser Scanning Microscope. The permeation of P2200 was also observed through cell wall without stacking to cell wall. The effective growth inhibition of bacteria was assumed to be the prevension of delivery of nutrition through cell wall.

INTRODUCTION

Chitosan, a mucopolysaccharide of $\beta 1,4$ linkage and deacetylated form of chitin, has been reported to inhibit bacterial growth(1). Several mechanisms were proposed for the antimicrobial activity by chitosan. One of major proposals is the reduction of bacterial metabolism by stacking of chitosan molecules to bacterial cell wall(2) . Another one is blocking of description to RNA from DNA by adsorption of penetrated chitosan to DNA molecules(1). In this mechanism, chitosan must be hydrolyzed during fermentation to the molecular weight less than around 5,000 which is easy to permeate into the cell, because chitosanases are contained in the fermentation medium, generally. We intended, in this study, to clear these mechanisms by applying chitosan oligomers of various molecular weights.

Chitosan

M. Kamachi · A. Nakamura (Eds)
New Macromolecular Architecture and Functions
Proceedings of the OUMS '95 Toyonaka, Osaka, Japan, 2-5 June, 1995
© Springer-Verlag Berlin Heidelberg 1996

Chitosan oligomers of various molecular weights were prepared by applying nitrous acid degradation of chitosan and FITC(Fluorescencein isothiocyanate) labeled chitosan oligomers were also prepared by direct modification. An antimicrobial activity was studies by chitosan oligomers at various concentration with use of five strains of bacteria(*E. coli, Bacillus cereus, Staphyrococcus aureus, Salmonella typhimurium, Enteropathogenic E. coli*).

EXPERIMENTAL

Chitosan: Chitosan was prepared from Queen crab shell through chitin by treating in concentrated sodium hydroxide aqueous solution at boiling temperature for 6 hours. The prepared chitosan was further treated by concentrated sodium hydroxide aqueous solution in a autoclave for 1 hour. The degree of deacetylation (DAC) was estimated from infrared transition spectrum according to Sannan et al(3).

Chitosan oligomers: Chitosan oligomers were prepared by depolymerization of chitosan with nitrous acid as shown in Scheme 1. Briefly, to a chitosan (DAC 73%) solution in 10%(w/v) acetic acid aqueous solution, 10% of aqueous nitrous acid (0.3 mole equivalent for amino group) was added dropwisely for 30 min. at ice cold temperature under stirring and the reaction mixture was stood for another 10 hours at 4°C to complete the degradation reaction. A 2 mole equivalent sodium borohydrate was added at 4°C to reduce the 2,5-anhydromannnose terminal to 2,5-anhydromannitol group under vigorous stirring overnight. The fractionation of reaction mixture was achieved either by ultra filtration or selective precipitation by poor solvent such as methanol following to concentration of supernatant for 1/3 of volume. Each precipitate was airdried after washing with methanol and followed by estimation of average molecular weight with Shimadzu Liquid Chromatography System JC6A-RID6A-CR6A and the degree of deacetylation through IR spectrum using Horiba SD-20 FT-IR spectrophotometer.

Preparation of FITC labeled chitosan oligomer: A 0.2g of P9300 or P2200 was dissolved in 20ml of H_2O:ethanol=1:1 mixed solution at pH 9.0 and 0.04% of Fluorescein Isothiocyanate(FITC) was reacted with chitosan oligomer at ice cold temperature under stirring overnight. FITC-chitosan oligomer was precipitated with ethanol and air-dried after extensive rinse with ethanol. The amount of incorporated FITC was estimated by fluorometry using Hitachi Fluorescence Spectrphotometer 650-60.

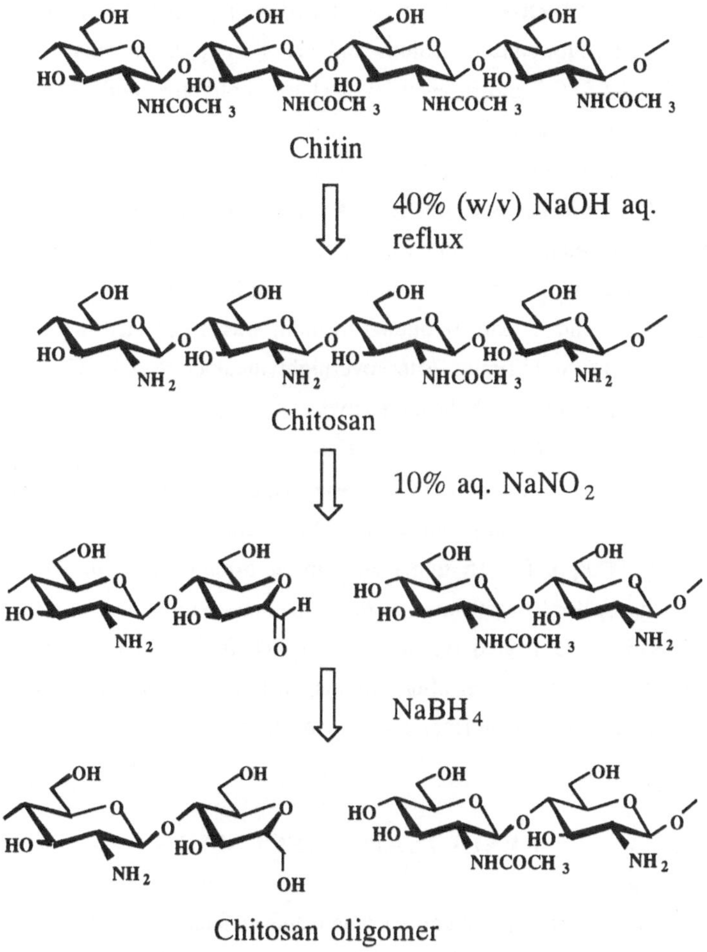

Chitin

40% (w/v) NaOH aq.
reflux

Chitosan

10% aq. NaNO$_2$

NaBH$_4$

Chitosan oligomer

Scheme 1 Synthetic route of chitosan oligomer by nitrous acid degradation.

Measurement of turbidity: *E. coli* was cultured in the Vogel-Bonner minimal medium for two times to adapt for culture medium and 2% of them was inoculated to the medium containing 0.01, 0.05, 0.1 and 0.5%(w/v) of chitosan oligomer under shaken cultivation at 30°C for 24 hours. During incubation, turbidity of the medium was measured at 610 nm on every 2 hours which was a indicator of cell growth.

Measurement of viable cell count: 2% of culture medium adapted *E. coli* was inoculated to the medium containing various amount of chitosan oligomer. The culture medium was withdrawn every 4 hours followed by dilution with 0.85% of the physiological saline and a part of diluted medium was suspended in the nutrient agar medium followed by solidifying. After overnight incubation at 28°C, the number of viable colonies was counted with the naked eye.

Fluorescence observation: 5% of culture medium adapting *E. coli* was inoculated in the medium containing 0.01% of FITC-chitosan oligomer under shaken culture at 30 °C for 24 hours in the dark. *E. coli* cultured with FITC-chitosan oligomer was collected by centrifugation (3,000rpm for 15 min.) and rinsed extensively with the physiological saline. The aggregated organisms were dispersed with ultra sonication in the physiological saline and loaded on the fluorescence free slide glass followed by air drying. Then cover glass was set after treatment with 50% aqueous glycerol containing sodium azide and localization of fluorescence was observed by Confocal Laser Scanning Microscope.

RESULTSAND DISCUSSION

The fractionation of chitosan oligomers was achieved successfully to obtain major two fractions. One (P9300) was the 9300 of weight average molecular weight (Mw), 1.80 of molecular weight distribution (Mw/Mn) and 51% of degree of deacetylation (DAC), and th other(P2200) was 2200 of Mw, 2.21 of Mw/Mn and 54% of DAC.

Among two major fractions of chitosan oligomers preparing by nitrous acid degradation of chitosan, a significant growth inhibitions for *E.coli, Enteropathogenic E.coli* and *Staphyrococcus aureus* were observed with chitosan oligomer of P9300
even at low concentration as shown in Figures 1-3. Chitosan effect was not so sensitive for *Salmonella typhimurium* and *Bacillus cereus* as shown in Figure 4 and Figure 5, respectively.

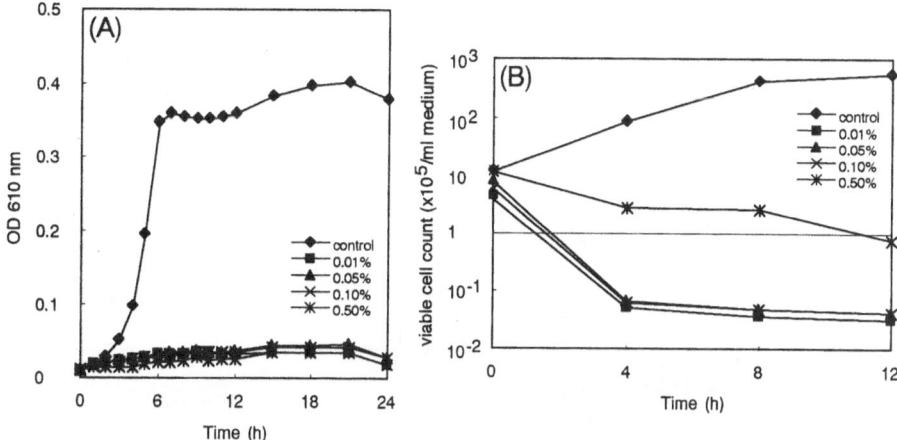

Figure 1 Growth curve(A) and viable cell count(B) of *E.coli* obtained
after incubation in the Vogel-Boner medium containing
various concentration of chitosan oligomers (P9300)

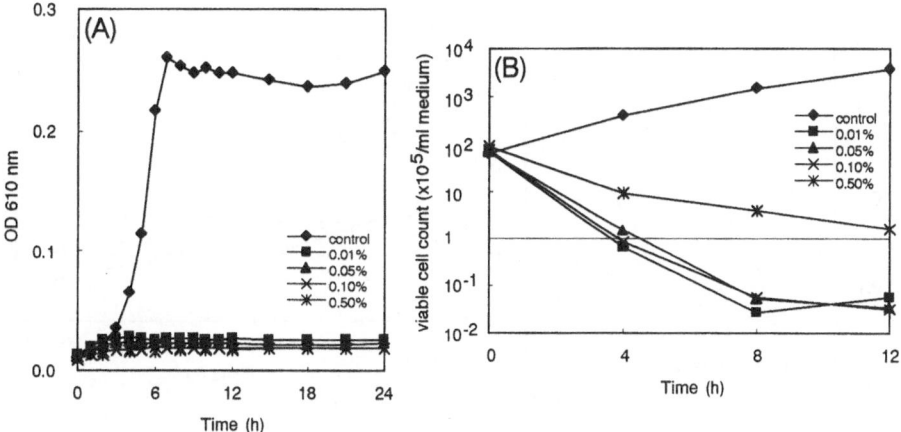

Figure 2 Growth curve(A) and viable cell count(B) of ETEC obtained after incubation
in the Vogel-Boner medium containing various concentration
of chitosan oligomers (P9300)

204

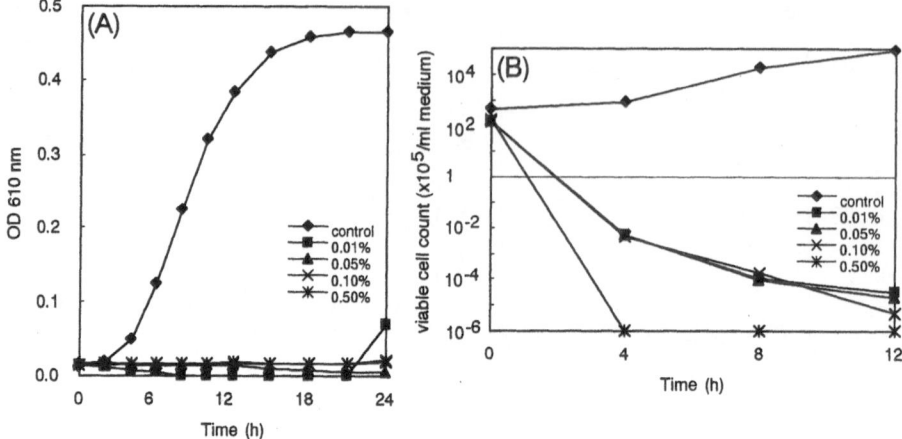

Figure 3 Growth curve(A) and viable cell count(B) of *St.aureus* obtained
after incubation in the Vogel-Boner medium containing
various concentration of chitosan oligomers (P9300)

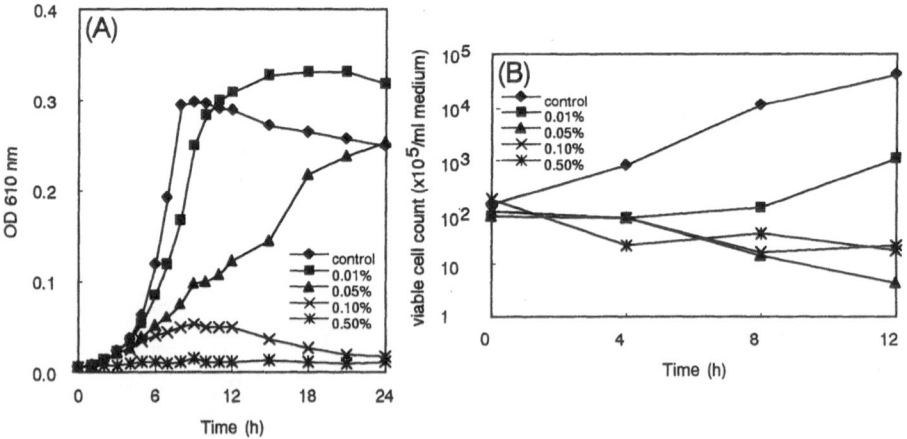

Figure 4 Growth curve(A) and viable cell count(B) of *Sa.typhimurium* obtained
after incubation in the Vogel-Boner medium containing
various concentration of chitosan oligomers (P9300)

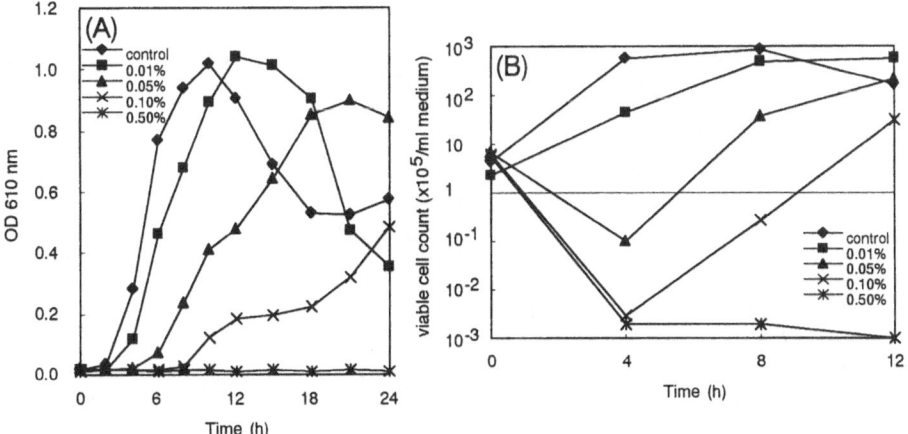

Figure 5 Growth curve(A) and viable cell count (B) of *B.cereus* obtained
after incubation in the Vogel-Boner medium containing
various concentration of chitosan oligomers (MW.9300)

But a slight acceleration of growth was observed for *E.coli* and *St.aureus*, when chitosan oligomer of low molecular weight (P2200) was applied as shown in Figures 6 and 7, respectively. The accumulation of fluorescence was found only for the cell wall of *E. coli*, when FITC labeled chitosan oligomer of higher molecular weight fraction (P9300) was added to fermentation medium as shown in Figure 8, whereas fluorescence was observed at the inside of the cell with use of FITC labeled P2200 as shown in Figure 9.

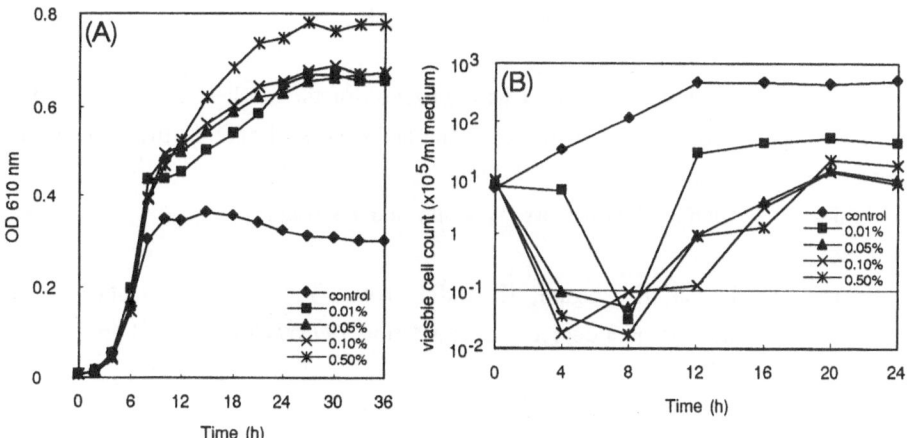

Figure 6 Growth curve(A) and viable cell count(B) of *E.coli* obtained
after incubation in the Vogel-Boner medium containing
various concentration of chitosan oligomers (P2200)

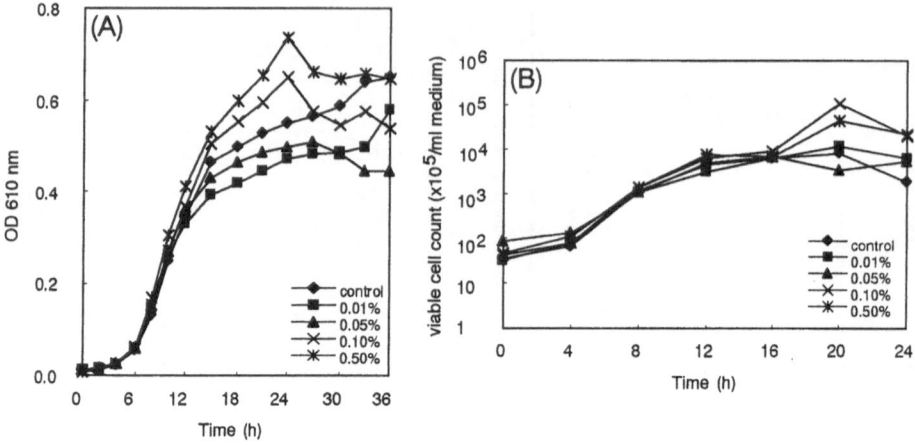

Figure 7 Growth curve(A) and viable cell count(B) of *St.aureus* obtained
after incubation in the Vogel-Boner medium containing
various concentration of chitosan oligomers (P2200)

Thus, permeated chitosan oligomer was not suggested to block the transcription from DNA, but rather to suppress the metabolic activity of the bacteria by blocking of nutrient permeation through cell wall in the case of P9300. As these fluorescence study corresponds closely with results of antimicrobial activity by chitosan oligomers, the antimicrobial activity of chitosan seems to be caused mainly by the block of nutrition supply through the cell wall of bacteria.

REFERENCES

1. Kendra, D.F., Hadwiger, L.A., Characterization of the smallest chitosan oligomer that is maximally antifungal to Fusarium solani and elicites Pisantin formation in pisum sativum. *Exp. Mycology*, **8**, 276-281(1984).

2. Uchida, Y., Antimicrobial activity by chitin and chitosan, *Food Chemical*, **2**, 22-29(1988).

3. Sannan, T., Kurita, K., Ogura, K. and Iwakura, Y., Studies on chitin: 7. I.r. spectroscopic determination of degree of deacetylation, *POLYMER*, **19**, 458-459(1978).

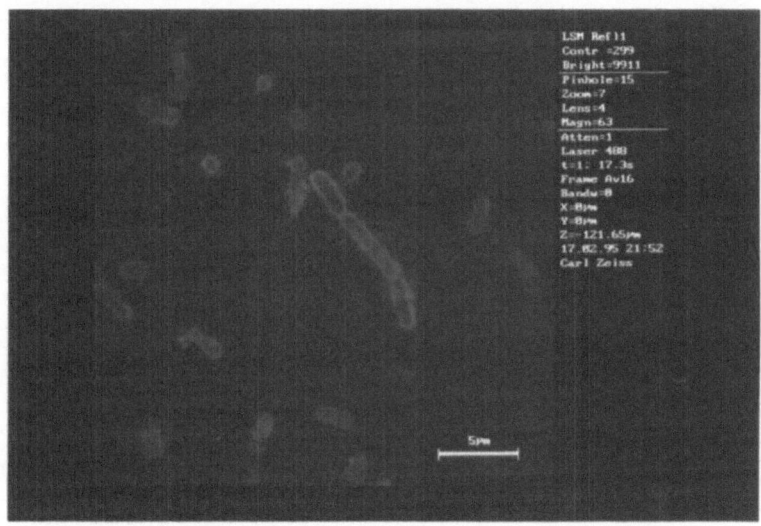

Figure 8 Fluorescein micrographs of E.coli stacked
FITC-labeled chitosan oligomers (P9300)

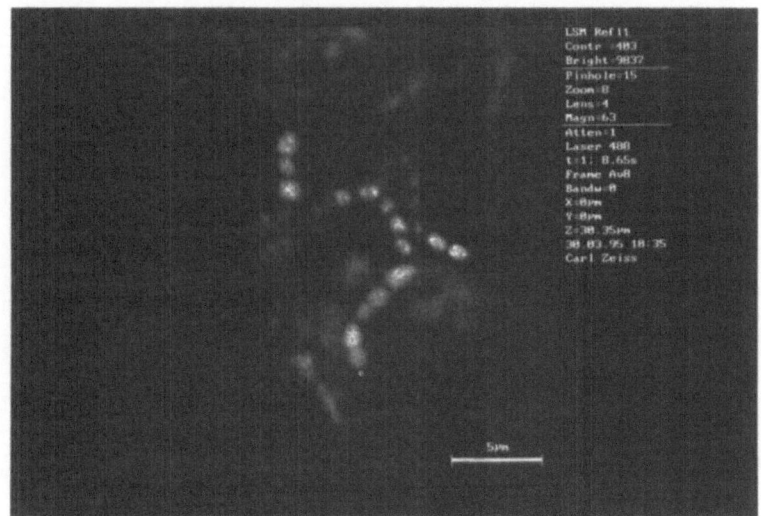

Figure 9 Fluorescein micrographs of E.coli accumulated
FITC-labeled chitosan oligomers (P2200)